PIECES OF THE GLOBAL PUZZLE:

INTERNATIONAL APPROACHES TO ENVIRONMENTAL CONCERNS

PIECES OF THE GLOBAL PUZZLE

INTERNATIONAL APPROACHES TO ENVIRONMENTAL CONCERNS

ANNE M. BLACKBURN

FULCRUM
GOLDEN, COLORADO 1986

Book Design by Bob Schram
Cover Design by Patricia Sullivan

Library of Congress Cataloging in Publication Data

Blackburn, Anne M.,

Bibliography: p.
Includes index.
1. Pollution. 2. Environmental protection. 3. Environmental protection — United States. 4. Environmental impact analysis — United States. I. Blackburn, Anne M. Title. II. The Environmentalist.

TD174.P53 1986 363.7'3 86-7583
ISBN 1-55591-002-5

FULCRUM, INC.
GOLDEN, COLORADO

Dedicated with thanks and love to Jim, Carol, Barbara, and Jeanne for their support; to Bob for the chance; and to Anne, David, Ben, and Michael for their continuing guidance.

Table of Contents

Preface

For four years my husband and favorite colleague, Jim Aldrich, and two dear friends, David Hughes-Evans and his wife, Jeanne Leger, of London, England, worked together to produce *The Environmentalist,* an international journal published by Elsevier Sequoia, of Lausanne, Switzerland. Starting up TEN (the publisher's internal code name for the journal) was our response to many years of concern over the need for more and better communication among professionals working in all the different aspects of environment and development, and with decision-makers in politics, government, industry and education, as well as the broader public. All too often the scientific or technical data that was being developed remained locked in the jargon and publications of the particular field from which it emanated. We began TEN as our way of making that highly useful information available to wider audiences by finding experts from those fields who were willing, and indeed anxious, to reach persons interested in their findings but who lacked technical background.

There were great satisfactions in producing TEN. There were also great frustrations. Time and time again a paper that would have made a magnificent companion to another would appear several months after the first had gone to press. Thus we were never able to fully orchestrate the information that came to us in ways that could optimize our readers' chances of understanding the individual problems discussed by one author, the inter-relationships of those particular issues to other societal concerns, and the cumulative importance of these issues to the long-term well-being of the United States and other nations of the world.

Thus, having the opportunity to prepare *Pieces of the Global Puzzle* is a dream come true. With the cooperation of my publisher, Fulcrum, and Elsevier Sequoia, the former owner of TEN, I have selected excerpts from papers and articles published in the 16 issues that appeared in 1981 through 1984, and supplemented these with personal observations and information from still other sources.

The book begins by sharing the views of several experts on the nature of global environmental pollution and natural resource problems, and the different ways these problems are affecting other societal concerns such as economic vitality and political stability. In the second section, the ways that four of the individual environmental problems are manifesting themselves in the United States are explored in greater detail. The final section investigates a number of innovative responses that global environmental issues have stimulated. My own observations are from one who has both worked intensively with and been a steady observer of the responses of United States local, state and national governments, and the international community, to steadily increasing pressures from environmental problems and the host of other global issues with which these environmental challenges interact. Updates on the efforts to resolve problems cited in the papers have been included wherever possible. A complete list of contacts is provided for those who may wish to learn more about any of the individual programs and projects that are mentioned.

Anne M. Blackburn
Wayland,Massachusetts
January, 1986

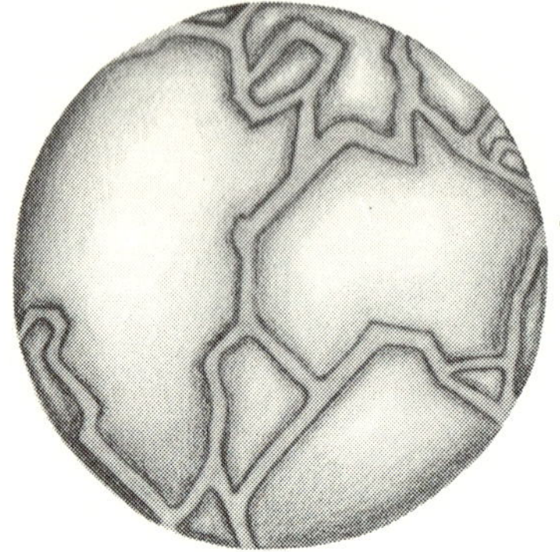

PART ONE

"This here young lady," said the Gryphon, "she wants for to know your history, she do."

"I'll tell it her," said the Mock Turtle in a deep, hollow tone: "Sit down, both of you, and don't speak a word till I've finished."

Alice's Adventures in Wonderland
Lewis Carroll

Understanding the New Global Puzzle

Thoughts from Environmental Leaders

Studying maps and working jigsaw puzzles have always been great ways to learn how things go together. Puzzles of the United States, no doubt, helped each of us develop a better sense of where we lived in relation to other parts of the country. The puzzles probably raised our curiosity about those places and led to improved retention when we began our study of them. Puzzles of the world, likewise, led to an improved understanding of where the United States was located in relation to other nations, how its size compared, with whom it shared its national borders and oceans, which other nations were in similar geographic locations around the globe. Creating this basic framework in our minds helped the rest of the facts and figures we would later acquire about world history and geography make sense. It helped open us to why different cultures and political systems might have evolved as they have. These experiences were some of the earliest steps we, as individuals, took toward understanding that we are citizens of many places — our homes, our local community or city, our state, our region, our nation, our continent, our hemisphere, and, indeed, our particular planet.

But the word puzzle also has another connotation — that of an as yet unresolved question or problem. And that aspect, too, is relevant to the issues that this book is exploring. These pages contain many, many examples of emerging challenges — how to stop the damage to global forests from acid rain while still providing book is exploring. These pages contain many, many examples of emerging challenges — how to stop the damage to global forests from acid rain while still providing needed electrical energy and transportation; how to control the use and disposal of toxic chemical in ways that retain the helpful work performed by these substances but also protect public health and natural resources; how to feed a growing global population and, at the same time, maintain the productivity of soils around the world; how to protect coastal fisheries resources while benefiting from the opportunities implicit in tourism; how to preserve a rich genetic pool; and how to protect all of the resources — natural systems and humans — from the impacts of war, especially nuclear war. These represent but a few of the questions addressed here. These issues, like many other political, social and economic needs currently vying for public attention, seem bewilderingly complicated, and almost unfairly insistent on challenging the existing systems we have developed to solve our problems and to interact as nations.

It is as though some mischievous spirit filched our old, familiar and comfortable puzzle of the world, based on national boundaries, and replaced it with a whole new configuration of puzzle pieces that we must now figure out how to put together. The global environmental issues examined in this book represent important parts of this new puzzle, but they are not the total picture by any means. Everywhere we turn we see multinational industrial developments; ever more complex import and export relationships; dependencies of developed world food supplies, medicines and manufactured products upon Third World resources; and worldwide transfers of investment capital. This new reality of global interdependence — of sovereign nations linked together in more and more complex ways — is what the challenging new global puzzle is all about. And thus, the global environmental problems presented here serve a dual purpose. They are important phenomena in and of themselves and they are also usefully visible examples of the types of connections that are linking together the community of nations which share this planet.

Recognizing the three main characteristics of these global environmental issues may help us understand the changing nature of the emerging links among nations. Transboundary and multi-disciplinary by nature, the problems ignore political borders within and between nations, and cannot be resolved without cooperation among many types of scientists, technical experts, political leaders — not to mention the long-term support of the general public. These issues, in the environmental vernacular, are also multimedia: resolution of the problems associated with any one may have potential impacts on other environmental and societal concerns. Thus few, if any, simple solutions exist.

The pages in this first section explore our improving understanding about how the planet's natural systems and resources function and interact. They also demonstrate that when environmental systems are overstressed, and their capacity to function is weakened or destroyed, the impacts radiate out — sometimes adding unrest and instability within individual countries, sometimes straining relations between nations, and sometimes affecting aspects of the world economy.

Like the strands of a spiderweb suddenly visible in the sunlight, we see these environmental problems as representative of the new global patterns that have been superimposed over traditional political boundaries. The issues raised by these problems help us understand why long-lived disputes between nations, as well as alliances based solely on similar cultures, shared political philosophies or common levels of development may now be less appropriate and may even be totally obsolete. Each interconnected environmental issue that we examine helps us understand the dramatic changes that are part of and made necessary by this new global puzzle of increasingly interdependent sovereign nations.

1. Conservation Strategies: for a Changing World

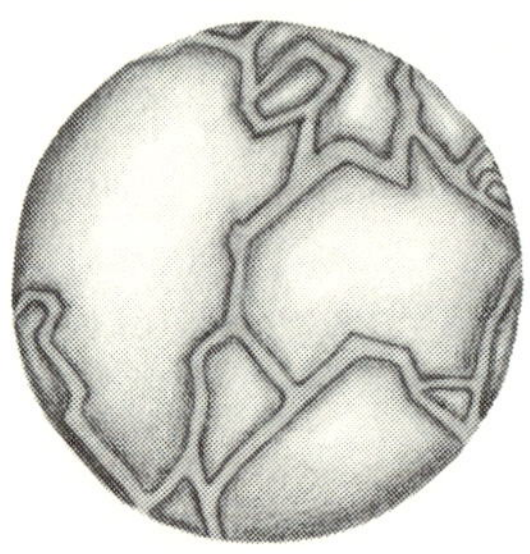

Maurice F. Strong
Member,
World Commission
on Environment
and Development
Published in 1981
Volume 1, Issue 4
The Environmentalist

The first contribution is from Maurice F. Strong, who was selected as lead spokesperson because of his extraordinary career. He has been a leading North American businessman; the government-appointed head of Petro-Can, Canada's national oil company; and the founder of the private-sector International Energy Development Corporation, which helped the developing world meet its oil and energy needs. In 1983 he was made chairman of the Canada Development Investment Corporation, which looks after that country's nationalized industries, while he continued as director of the vast American private conglomerate, Tosco. Mr. Strong was Secretary General of the United Nations Conference on the Human Environment (Stockholm, 1972), and served as first executive director of the United Nations Environment Programs established as the implementing secretariat for the recommendations approved at Stockholm. He helped establish the International Institute for Environmental Development, a non-governmental organization which has become a primary force on the global environmental and human-needs front. Mr. Strong is also one of the 23 world leaders appointed to the World Commission on Environment and Development, which has the responsibility of reporting to the United Nations, in 1987, on The Environmental Perspective to the Year 2000 and Beyond. Late in 1984 Mr. Strong responded to the request for his services at the United Nations Office of Emergency Operations in Africa, where he has chosen to serve as executive officer without pay. This office was established to coordinate short-term assistance to the 50 African nations suffering from famine. We share his thoughts as expressed in the Third World Conservation lecture, held by the World Wildlife Fund/U.K., October 2, 1983. The lecture was published in 1984 in Volume 1, Issue 4 of *The Environmentalist.*

1 I would like to explore with you the changing world in which we who are committed to conservation of the Earth's resources must pursue our goals in the period ahead.

First let us acknowledge that the very concept of conservation has itself changed to a significant degree in recent times. In the past the emphasis was often on preservation, on resistance to development and other forms of human intervention. Now the emphasis has, in my view, properly shifted to a more positive and dynamic concept of the care, stewardship, and wise use of resources, to one which recognizes that natural resources themselves are in a constant state of change, that human interventions are a necessary part of that process, and that the primary role of the conservationist is to ensure that the process is managed in such a way as to prevent destruction of the resources, and to maintain the integrity and the beauty of the ecosystems of which the resources are a part. There is thus a direct and inevitable link between conservation and human intervention, particularly that which has economic and social growth as its objective, which we commonly these days refer to as development.

In little more than a generation, conservation has moved from the fringes of human activity and public awareness into the center of the world stage. In the past, conservation was largely a preoccupation of scientists, specialists and avid nature-lovers interested primarily in the preservation of endangered species of animals, wilderness areas and other natural areas of particular beauty or interest. Today, conservation-related issues have to be on everyone's list of most important public concerns in virtually every country of the world. Indeed, I would contend that the combination of continued population growth and the drive to improve the standards of life for the majority of people who live in the developing countries makes conservation central to the very survival and well-being of all human life on our planet.

As our concepts of conservation have been changing, the world to which we must relate these concepts has been experiencing profound changes which, in turn, have transformed not only the objective factors to which the substance of our conservation efforts must be directed, but also the social, political and economic conditions in which we must pursue them. The United Nations Conference on the Human Environment, held in Stockholm in 1972, was in many ways a watershed in the development of our awareness of the importance of conservation. It was the first time that governments of the world came together at the highest levels to recognize that the same processes of industrialization and urbanization which had produced great economic benefits, principally for peoples of the industrialized world, were giving rise to potentially dangerous consequences for human health and well-being, and were destroying the renewable natural resources and major ecosystems on which human life and well-being depend. Although the initiative for the conference came from the industrialized countries, who were concerned primarily by their own problems of pollution and deterioration of the quality of urban life, the developing countries made it clear that their principal environmental problems derived from their poverty and underdevelopment. They made the point that their primary commitment must continue to be to development as the means for enabling their people to realize their aspirations for a better life. These concerns of the developing countries played an important part in shaping the results of the Stockholm Conference as well as the mission and the activities of the United Nations Environment Program which was established as a result of the conference.

The Stockholm Conference produced a worldwide wave of public interest and concern, and led to a broad range of new initiatives on the part of governments, international organizations, non-governmental agencies and citizens groups. Particularly notable was the action of developing countries, most of which set up some form of environmental agency or ministry, and initiated national environmental programs and legislation.

Since Stockholm there has been progress, some of it notable, in dealing with virtually every environmental concern that was on the Stockholm agenda. The good news is that experience has demonstrated that most issues can be resolved positively — that the need to conserve resources and protect the environment can be reconciled with economic growth and development. The bad news is that, in most instances we are still doing much less than we can do, and must do, to ensure the maintenance of a secure and healthy environment for human life on planet Earth — that despite the progress made on many fronts, on the whole there has been a deterioration of the natural resource base of our planet since the Conference on the Human Environment in Stockholm.

Although destruction of tropical forests has been stopped or slowed down in some places, overall the rate of destruction continues at a pace which threatens the very existence of these invaluable and irreplaceable ecosystems. Programs of reforestation and recovery of lands that had reverted to desert have increased significantly; yet far more productive land continues to be lost through careless and destructive exploitation than is being reclaimed by all these programs. Some notable successes have been recorded in saving endangered species of animals and plants, and more attention is now being focused on preserving genetic stock; nevertheless, more species of animals and plants are in danger of extinction today than ever, including some that are vital to maintaining world food production.

In industrialized countries there has been notable progress in the control of oil and water pollution, but some of these countries have at the same time permitted their industries to transfer production facilities and products to developing countries under conditions which pose risks to human health and the environment that would not be allowed at home. There has been little progress on enforceable international agreements for monitoring and controlling the trade of toxic chemicals or their use. And there is a particularly dangerous increase in the exposure of Third World peoples to the hazards of chemical products, including those already banned as dangerous in their countries of origin. Indeed, there is a mounting danger that, as one observer put it recently, the Third World will become the dustbin for the excess and rejected products of the chemical producers of the industrialized world.

The atmosphere is cleaner today in many of the cities of the industrialized world, as, for example, London, New York, Tokyo; but it is worse, and continuing to worsen to levels which endanger the health of residents, in many cities of the developing world, including Mexico City, Santiago, Cairo and Calcutta. Rivers and lakes have become cleaner in some parts of the industrialized world, but in developing countries many rivers are becoming progressively more polluted, posing growing hazards to the health of millions of people.

The late Barbara Ward — Lady Jackson — who, with the late Professor Rene Dubos, authored *Only One Earth*, the book which so inspired and guided the Stockholm Conference, pointed to the non-availability of clean water and sanitation as the principal single threat to the health and well-being of people in the developing world, particularly children. Since then, there has been growing recognition of the importance of this issue, as evidenced by the decision in 1980 of the United Nations General Assembly to designate the 1980s as the "International Drinking Water Supply and Sanitation Decade." The United Nations estimated that to provide clean water and adequate sanitation for all the developing world by 1990 would require some $25 million a day in development assistance (the world spends about 10 times that amount a day on cigarettes, and $1,400 million a day on armaments). Yet there has been little or no increase in the aggregate resources being made available to meet this need, and little prospect that the needed resources will be available in the foreseeable future.

Acid rain was identified at Stockholm as a threat to freshwater lakes and forests. This issue has become an even more acute one today, particularly in Canada, the United States and the Scandinavian countries, where it is escalating into a major source of conflict amongst otherwise friendly nations.

The oceans of the world seem to have shown more resilience against oil spills and other forms of contamination than was apparent at Stockholm. And there has been a noticeable increase in international cooperation for protection of the marine environment, as exemplified by the Regional Seas Program of the UN Environment Program and by completion of international conventions controlling ocean dumping of toxic chemicals. These positive developments have not solved the problems of ocean pollution, but they have given us hope that the problems can be solved and have establish the means for doing so.

The tough issues such as long-term control of land-based sources of pollution and dumping of radioactive materials and other toxic waste remain largely unresolved, and progress toward resolving them has clearly slowed down, if not slipped back. The lack of universal approval of the Law of the Sea Treaty threatens to frustrate its objectives and create a whole new set of risks and uncertainties for peaceful management and protection of this 70 percent of the Earth's area. It also raises the ominous prospect that the oceans may become a new source of division and conflict amongst nation states and between rich and poor.

Commendable and enlightened cooperation by the Antarctic Treaty powers has thus far spared the Antarctic from the principal threats to its highly sensitive and critically important environment, and provided an impressive body of research and knowledge to guide future decisions. But there are disturbing signs that the treaty powers are quietly preparing the way for development of the Antarctic, particularly petroleum exploration. This would represent a major challenge to the concept that the Antarctic is and should remain an intrinsic part of the international commons — part of the common heritage of all mankind, rather than the preserve of a few privileged nations. I fear, and think it likely, that the Antarctic may emerge in the period ahead as a major new arena for international conflict, of which conservation of this vital ecosystem will be a key element.

Although there is comfort for many in the fact that the growth in the use of nuclear power as an energy source has been much slower than was predicted, and feared by many, at the time of Stockholm, this has been more than offset by the alarming expansion of nuclear weaponry. The capacity for nuclear destruction is greater today than it has ever been, and escalating tensions and conflicts throughout the world have deepened the threat that this ultimate sword of Damocles may yet be used, with all its devastating consequences for the very survival of the human species.

Since 1972 the rate of world population growth has slowed down and many developing countries have committed themselves to policies and programs designed to limit population growth. But there are still almost 1,000 million people more on Earth now than in 1972, and population will continue to grow, largely in the developing world, for the foreseeable future. In the meantime, per capita use of resources has also grown. The result is that pressures on the Earth's resources and the biosphere are still increasing and will continue to grow. These pressures will be concentrated more and more in developing countries, where population growth is greatest and per capita resource use must increase most.

One of the most important perceptions to emerge from the Stockholm Conference was recognition by the industrialized countries of: a) the importance to the future of the entire world of conserving the natural resources of the developing countries; and b) the compelling need to increase the flow of resources, both financial and technical, to these countries, to enable them to develop in ways which do not destroy or undermine the resource base on which future development depends. Sustainable development is not an option for the developing countries; it is a stark necessity. But people faced with the struggle for day-to-day survival cannot be expected to give priority to preserving resources for tomorrow when they have to make a choice.

The plight of the developing countries has become much more acute and their prospects less promising.

Worldwide recession, increased energy costs, mounting burdens of costly debt and the lowest commodity prices for some 30 years, have crippled the fragile economies of many developing countries, including some of the strongest and most dynamic, like Brazil and Mexico. The effect on the poorest countries has been even more devastating, and their chances of recovery without substantially higher levels of international assistance are grim indeed.

Yet assistance from the industrialized countries is not growing, and in real terms has in fact been diminishing. Although it is gratifying to note, from a conservation point of view, that more funds are flowing into conservation-related activities — reforestation, fuelwood plantations, measures for forest and soil conservation, and increased attention to the environmental impacts of major projects — this is largely attributable to the transfer of resources from other areas rather than to an increase in aggregate resource flows.

Despite the tremendous pressures which these problems have generated for leaders of developing countries, it is encouraging to report a growth in concern and awareness of the importance of conservation in these countries, which is nothing short of remarkable in the circumstances. This is manifest at the highest levels of government in the enactment of legislation and the initiation of policies on environmental protection and conservation of resources by many developing countries, but it is also manifest in a surprisingly high degree of citizen interest. In Zambia recently I met with a group of farmers and rural development officers from throughout the country concerned with conservation of soil and water in their areas. In Indonesia I was encouraged to find a vigorous and active citizens' movement with some 400 organizations cooperating closely with the government, largely through the enlightened leadership of Dr. Emil Salim. Grassroots citizen interest and concern has been manifest in many instances in India, including the famous chupka movement in which local people literally chained themselves to trees to prevent the trees from being cut down. A businessman from Latin America recently complained to me that it was impossible to initiate any new project without the intervention of con–cerned citizens. And there has been a substantial increase in the number of national parks and nature reserves established by developing countries.

But, as encouraging as this is, it is still too little, and in some cases too late, to stop the massive destruction of natural resources which continues to undermine the ability for many developing countries to sustain the kind of development they require to meet the future needs and aspirations of their people. The grim evidence of this was well documented in the *World Conservation Strategy* prepared by the International Union for Conservation of Nature and Natural Resources in cooperation with the UN Environment Program and the World Wildlife Fund. It points out:

• More than half of India suffers from some form of soil degradation; out of a total of 3.3 million square kilometers, 1.4 million are subject to increased soil loss, while an additional 270,000 are being degraded by floods, salinity and alkalinity. An estimated 6,000 million tons of soil are lost every year from 800,000 square kilometers alone; with them go more than 6 million tons of nutrients — more than the amount that is applied in the form of fertilizers.

• Hundreds of millions of rural people in developing countries, including 500 million malnourished and 800 million destitute, are compelled to destroy the resources necessary to free themselves from starvation and poverty.

• For the 500 million people who are malnourished, or the 1,500 million people whose only fuel is wood, dung or crop waste, or the almost 800 million people with incomes of $50 or less a year — for all these people conservation is the only thing between them and, at best, abject misery, or, at worst, death.

• Fuel wood is now so scarce in the Gambia that gathering it takes 360 woman-days a year per family.

Even more dramatic is the evidence that presents itself to observers traveling in the developing world. I am sure some of you have seen the devastating effect of floods in the plains of India and Pakistan resulting from deforestation and disruption of the watersheds of the Himalayan region in which the great rivers of that area rise. And in the valleys of this region, the pitiful attempts of peasants to glean a living from small plots on mountain slopes denuded of soil, and scarred by the clefts and ditches created by untamed stormwater runoffs (while facing constant threats from landslides and avalanches) testify to the horrendous human consequences of ecologically unsound development. Similarly, in the valleys of Burundi in Central Africa the growing population has led to deforestation of the hillsides, overintensive cultivation of areas that would not sustain it, and massive loss of soil, all of which threaten the prospects of a large part of the population of this small country. Flying over the Tana River in Kenya, one can see the devastating consequences of deforestation along the river banks, turning the river into a river of mud, destroying the fishing which once provided an important source of livelihood to people of the area. Then as it flows into the Indian Ocean it creates a vast arc of silted water around the estuary and spreads a layer of silt over the beautiful white sand beaches, thus degrading one of Kenya's principal tourist areas. And in the Sahel region of Western Africa and many other desert areas of that continent, one can see cultivated fields and villages being gradually, but relentlessly, overtaken by the spread of the desert. This process of desertification results from the removal of trees and plants by poor people in the continuous search for food and fuel, not only from ecologically unsound cultivation practices, but also all accentuated by the occurrence of the droughts which periodically affect those regions.

The World Conservation Strategy fortunately does not stop at documenting the urgent necessity of addressing the major conservation issues which affect the future of the human community and particularly the peoples of the developing world; it provides a clear and practical framework for doing this. But, of course, most of these issues have to be tackled on the national level.

And since the promulgation of the World Conservation Strategy in 1980, the primary emphasis of the international conservation community has, properly, shifted to encouraging and supporting the development of National Conservation Strategies.

Today some 30 countries have, or are in the process of developing, such national strategies. This includes such developing countries as India, Nepal, Fiji, Indonesia and Zambia. The purpose of the National Conservation Strategies is to provide, at the national level, a means for integrating conservation measures into national development and resource management planning, policies and action.

Another recent initiative of the International Union for the Conservation of Nature and Natural Resources, supported by the World Wildlife Fund, has resulted in the establishment of the Conservation for Development Center, which is designed to provide practical assistance and advice on the conservation aspects of development planning and projects. Its task is to draw upon the resources and experience of IUCN's extensive network of member organizations, its specialized commissions and conservation professionals, as well as development assistance agencies and international organizations. It has also established close cooperation with the London-based International Institute for Environment and Development. The Conservation for Development Center is playing a leading role in helping developing countries to produce their own National Conservation Strategies and to carry out a variety of projects within the framework of these strategies.

Thus, the world conservation community is responding to the dire threats to the future of the human community posed by the destruction and degradation of the renewable resources of our planet. Conservationists must be in the vanguard of the movement to stop the vicious cycle of destruction of our renewable resource base. It cannot be done without us. But neither can it be done solely by conservationists.

We welcome the changes that have been made in the past two decades as environment has emerged as a major public concern, and this has been accompanied by growing awareness of conservation and its relationship to economic growth and development. But this change has not gone far enough. Our present modes of growth and development and the lifestyles which depend on them are simply not sustainable. The care and maintenance of planet Earth, and the future survival and well-being of its inhabitants, require that we adopt patterns of growth and ways of life which permit nature to complete the cycles by which it renews on a continuing basis the resources which people need and depend on for the basic elements of life — food, shelter, fuel and clean water.

We cannot continue to live on this planet by running down its resources, by using renewable resources — soil, water, plant and animal life — on a basis which makes them non-renewable. This will make planet Earth a failing enterprise, destined for bankruptcy, just as surely as a business which does not provide for amortization and maintenance of its plant and equipment and operates only by running down its capital.

We can put the business of planet Earth on a sound track only when we have brought growth and development into balance with the renewable resource base on which sustained growth and development depend. This requires a new approach to growth, in which a healthy balance is maintained between the renewable resource base, the capacity of the earth and water and atmosphere to absorb our waste products without unacceptable levels of damage — and the human activities which use and impinge on these resources. This new growth approach will require substantial changes in industrial practices — the adoption of more closed-circuit industrial systems, in which waste products are recycled or used — as well as less energy-intensive methods of production and transportation; more careful farming and forestry practices, which permit sustained yields on a continuing basis; and personal lifestyles that are less wasteful and more conserving of resources.

There has been some notable progress in these directions already. Indeed, it can be said that some of the elements of the new growth society are already in place.

For example the newest steel plants and oil refineries are vastly more efficient in energy use than their predecessors and they have drastically reduced emissions of pollutants to levels that are reasonably manageable. And many areas in the U.K., other parts of Europe, and even in parts of the developing world, like Hunza in the Himalayas, have demonstrated that ecologically sound agricultural practices can maintain both the natural beauty of rural areas and their productivity indefinitely.

These examples — and there are many more of them — make the point that sustainable lifestyles and patterns of development are feasible in both industrialized and rural societies. However, these examples are the exception more than the rule, and they apply more to the industrialized countries than to the developing world, where the unrelenting pressures of population growth and the struggle to meet basic needs are having a devastating effect on the renewable resource base.

The desperate economic plight of most developing countries, including some of the largest and potentially richest of them, is making it more and more difficult for developing countries to adopt conservation measures where they involve an immediate cost in exchange for a future benefit. And it is making it virtually impossible to provide people whose need for food and fuel-wood is causing widespread destruction of forests, soil and watersheds with alternatives which will meet their needs on a sustainable basis.

In many areas of the developing world, in which a fragile but sustainable balance has been achieved over thousands of years between human populations and the ecosystems which supported them, that balance has been disrupted. As a result, these areas are in the process of turning into vast wastelands, global slums whose inhabitants are doomed to abject poverty, their dreams of development and a better life shattered beyond redemption. What a paradox it is, what commentary it is on the way in which we set our priorities, that in this generation — the first in which we have developed an understanding of the ways in which our activities impact on the Earth's resources and life systems, and the first in which it is at least technically feasible to ensure that every person on this planet has access to the ingredients of a decent life — the stage is being set for one of the greatest human tragedies of all time. For I am quite convinced that the destruction of renewable resources in the developing world poses a threat as great to the human future as the prospect of nuclear war. And in many respects it is even more difficult to deal with. The fact that nuclear war presents a dramatic prospect of such unmitigated horror to all people provides in itself a continuing incentive to prevent it; and until the button which unleashes that horror is pressed, there is always the chance that it will not be pressed. By contrast, the destruction of living resources is like a cancer spreading quietly and relentlessly through the body of our planet. By the time its devastating consequences are manifest in depriving tens of millions of people of their means of life, it will be too late to cure.

Yet cures are available through alternatives that are often quite simple: improved means of cultivation and selection of the mixes of crops which will produce the best yields on a sustainable basis; application of the techniques of agroforestry which produce a more sustainable and productive balance between forestry and agriculture; replanting of trees so that exploitation of forests can be put on a renewable basis; reforestation of areas which have been denuded of their original tree cover but will still support new growth; establishment of fuelwood plantations. These are but a few. But all of these alternatives, to one degree or another, involve changes in the present habits and practices of people, particularly those who have been forced into patterns of life which are destructive of resources through the pressures of poverty and the imperatives of survival. The knowledge is available, the techniques are known, and the resources required to do this are well within our means (but a portion of the amounts being spent by governments today for military purposes).

The problem clearly is one of will. We speak of one world, of world community, of interdependence. And indeed we have developed a general, if somewhat vague, sense that somehow the fate of all of us on planet Earth is today inextricably linked as never before — that in the final analysis there can be no security, no sanctuary for any if there is not at least a minimum of security for all.

We have clearly not translated these concepts into realities. But we have made a beginning. We must take heart from our successes and build on them. Despite the many other issues crying for the attention of the public and our leaders, we must reinforce the voice of conservation. The World Conservation Strategy and the development of National Conservation Strategies have given us new instruments through which to create a constructive interaction between conservation and development. We must do everything possible to strengthen these instruments, and to use the many outlets we have to ensure that the conservation movement is making its full contribution to the emergence of the kind of world community in which all people may share in both the responsibilities and the benefits of life on our beautiful planet.

2. Sustainable Development: The Global Imperative

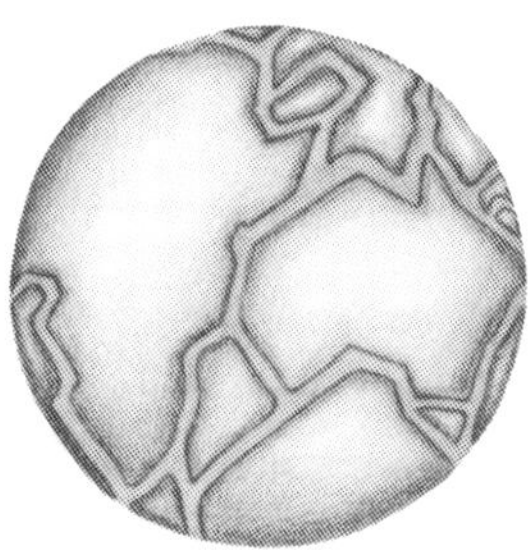

A.W. Clausen,
President
The World Bank
Published in 1982
Volume 2, Issue 1
The Environmentalist

Our second contribution is from A.W. Clausen, president of the World Bank, which is part of the system of international institutions set up after World War II to provide better means for international cooperation on issues of global concern. Owned by 141 nations, the World Bank lends funds to Third World nations to accelerate their economic development. Mr. Clausen assumed the presidency in July 1981, following a 32-year career with Bank of America and the Bank America Corporation, where, for the preceding 11 years he was president and chief executive officer of both institutions. He has also served as a director of many corporations and institutions, including Standard Oil of California and the U.S.-U.S.S.R. Trade and Economic Council. Mr. Clausen's remarks were given at a Washington, D.C., conference in 1981, where he delivered the Fairfield Osborn Memorial Lecture in Environmental Science. Mr. Clausen discusses sustainable development and the role of improving economic productivity in Third World nations, if that important goal is to be reached. The lecture was published in 1982 in Volume 2, Issue 1 of *The Environmentalist*..

2 Fairfield Osborn, in whose honor the Fair field Osborn Memorial Lecture in Environmental Science was established was, himself, a businessman —an investment broker, who was concerned about short-term economic development and also about its long-term sustainability. As founder and then president of The Conservation Foundation, a co-sponsor of this memorial

lecture, Osborn worked until his death in 1969 to arouse the concern of people everywhere to the "accumulated velocity with which (man) is destroying his own life sources." In his book, *Our Plundered Planet,* which appeared in 1948, Osborn wrote: "We are rushing forward unthinkingly through days of incredible accomplishment . . . and we have forgotten the Earth, forgotten it in the sense that we are failing to regard it as the source of our life." Fairfield Osborn insisted that the only kind of development that makes sense is development that can be sustained.

Beginning, then, from this basic premise, I'll make three main points: 1) that if our goal is sustainable development, our perspective must be global; 2) that human development must allow for continued economic growth, especially in the Third World, if it is to be sustainable; and 3) that sustainable development requires vigorous attention to resource management and the environment.

If our goal is sustainable development, then our perspective must be global

The conservation movement began in the industrial countries. But the industrial countries are linked together with the developing countries — more than is usually recognized. They are linked economically and they are also linked environmentally. This is obvious in the case of energy. The prospect of running down fossil fuel reserves (oil, gas and coal) has become a major world issue, with certain Third World countries crucial in deciding its outcome. Other mineral resources are finite, too. And, again, developing countries — major suppliers of many of the key minerals used by industrial countries — will have a say in determining how these resources are managed.

Sustainable population growth is a fundamental environmental concern. By the year 2000, the world's population is likely to exceed 6 billion, with nearly 5 billion people in just the developing countries and over half of them crowded into cities. Many of the environmental problems we tend to associate with industrial countries are also prevalent in developing countries. Urban air pollution, for example, is often worse in countries that can't yet afford even minimal controls.

But some environmental problems of world interest are concentrated in the developing countries — deforestation, for example. Primarily because of the growing need for firewood, Third World forests are being cut down 10 times faster than new ones are being planted. As a result, planet Earth has lost about a quarter of its closed-canopy forests over the last 20 years. The developing world also has soil problems. Deforestation has contributed to severe erosion in some parts of the developing world. In tropical areas, soils tend to be especially fragile and can quickly be rendered almost useless by improper agricultural practices. Environmental spoilation is an international cancer. It respects no boundaries. It erodes hard-won economic gains and thus the hopes of the poor.

Finally, sound development worldwide contributes to peace. And the world's conservation, in the most literal sense, may depend on peace.

In the early 1970s, when environmental concerns were approaching a peak in the industrial countries, these same concerns were met with considerable skepticism in the Third World, where intense aspirations for development are widespread. In many nations, economic growth is a matter of life or death for thousands of people on the margin of subsistence. Third World leaders were concerned that environmental measures not keep their peoples from fully realizing the benefits of economic growth. Since the industrial countries consume most of the world's natural resources, concerns in the rich countries about population growth in poorer countries seemed misplaced — even sinister — to some leaders in the developing countries. Many of the world's environmental problems increasingly depend on Third World cooperation for their solution. Success on the environmental front must involve the cooperation of all sectors of the international community.

Over the last decade, the World Bank has supported the goal of achieving sustainable development on a global scale. Owned by 141 member nations, the Bank lends to Third World countries in support of their long-term development objectives. It is entirely appropriate that we not only continue but expand our efforts to insure that improvements achieved in human living conditions today are improvements that can last until we reach tomorrow.

To be sustainable human development must Aaso include economic growth

Even in wealthy countries, environmental protection has, to some extent, been crowded aside by priority attention to economic difficulties. Yet the world's recent economic troubles have also been particularly severe for developing countries — and the need for economic growth in these countries is overwhelmingly compelling. Third World leaders are absolutely right to point out that poverty is the very worst pollution that faces us on Earth today. For example, only about a quarter of the people who live in developing countries have access to clean water.

In the Third World, disease typically takes up a tenth of a person's potentially productive time. Disease causes suffering, dampens initiative, disrupts education, and stunts physical and mental development. Poverty also puts severe —and often irreversible —strains on the natural environment. At survival levels, people are sometimes compelled to exploit their environment too intensively. Poverty has often resulted in long years of mismanagement of our natural resources, evidencing itself in overgrazing, erosion, denuded forests and surface-water pollution.

Our experience at the World Bank seems to indicate that it is much easier to deal with the negative environmental effects of development than with the negative environmental effects of pervasive and persistent poverty. The developing regions of the world aren't all characterized by severe poverty, of course. But aspirations for a better life — more schooling, economic opportunity and freedom of choice — are also powerful among the middle-income countries. In fact, aspirations for economic growth may be strongest in countries at moderate levels of income, i.e., those countries which have enjoyed some growth over the last generation. These hopes and aspirations are legitimate. They won't be easily denied — nor should they be denied.

I'm encouraged that environmental concerns in Third World countries have become much more widespread over the last decade. According to the World Environment Center, 122 developing countries now have environmental ministries or similar top-level agencies. That compares to only 11 countries in 1972, the year of the now historic United Nations Conference on the Human Environment. The People's Republic of China recently established an Office of Environmental Protection. Brazil's Ministry for the Environment, established in 1972 with three people, now has a staff of 200. Indonesia set up an environmental ministry in 1978 to deal with its enormous problems of water supply, deforestation, erosion and industrial pollution, and to oversee the environmental implications of its development activities. Two notable examples of countries that now consider environmental impacts much more carefully in the process of industrialization are Kuwait and Nigeria. Venezuela, too, has made much progress in recent years in its awareness of environmental issues and the need to take corrective and preventative measures.

It isn't that Third World countries are better able to afford environmental protection than they were 10 years ago. They continue to be hard-pressed by slow growth, high energy prices and high interest rates. But there is increasing awareness that environmental precautions are essential for continued economic development over the long run. Conservation is not a luxury that provides scenic parks in remote areas for people rich enough to visit such sites. And conservation is not just a "motherhood issue." Rather, the goal of economic growth itself dictates a serious and continuing concern for resource management. Lower-income countries also — perhaps especially — need to think and to plan with this longer term view in mind.

Agriculture is the major source of income and

growth in most developing countries. That's why agriculture and rural development is one of the top priorities at the World Bank. And continued agricultural productivity certainly requires protection of the natural environment. In Africa, for example, rain-fed agriculture will necessarily be the main source of increasing production for the foreseeable future. Yet the long-term potential for agriculture is being diminished — by overintensified cultivation and overgrazing.

In the industrial countries, we are accustomed to an adversary relationship between the goals of economic growth and environmental protection. But in the poorer developing countries, continued growth more often depends on environmental stewardship — and, conversely, a better environment more often than not depends on continued economic growth. And so the point is that, in the Third World context, the twin goals of development and sustainability can be allied — and must be allied. Thus our vision of a sustainable world must realistically include economic growth, especially for those peoples who haven't yet participated appreciably in the benefits of economic growth.

In order to be sustainable, development must include attention to resource management

There are two different types of resource management — actions to mitigate negative side-effects of development on the one hand, and positive actions to enhance the environment on the other. First let me review our experience at the Bank in taking precautionary measures against the negative side-effects of development.

The Bank is well-known for the thoroughness of its project preparation. And, for a decade now, the Bank has required, as part of project evaluation, that every project it finances be reviewed by a special environmental unit. As a matter of policy, we won 't finance a project that seriously compromises public health or safety; that causes severe or irreversible environmental deterioration; that displaces people without adequate provision for resettlement; or that has transnational environmental implications which are importantly negative. Our environmental experts have reviewed more than 2,000 projects and programs in developing countries since 1970. We have found that the cost of paying attention to environmental concerns has been much lower than many people expected when this procedure was first established. Nearly two-thirds of the projects reviewed have raised no serious health or environmental questions, and I'm pleased to say that it has been possible to incorporate adequate protective measures in all the projects we have financed during the past decade.

The cost of these environmental and health measures has proved not to place an unacceptable burden on our borrowing countries. And we have learned, as have many private corporations, that the cost tends to be lower the earlier that environmental problems are identified and handled. We are convinced that it is almost always less expensive to incorporate the environmental dimensions into project planning than to ignore them and pay the penalties at some future time. It costs a lot less, for instance, to protect the forested watershed above a new dam than to deal with a reservoir that fills with silt after it has been built. Similarly, the benefits of an irrigation project can be diminished if, for lack of proper planning, it leads to an increase of Schistosomiasis.

We are financing an increasing number of projects to rectify undesirable side-effects of past development. We are working with the Mexican authorities, for instance, on various environmental problems, including air and water pollution in Mexico City. In Kenya, we are financing a wildlife management project. We are also involved in helping Nepal deal with its dramatic problems of deforestation and soil erosion. In the Sahel region, we have financed a series of projects designed to halt the expansion of the desert and, more recently, to also reclaim desertified land by providing shelter belts of restored forest cover.

On the basis of the Bank's experience in trying to anticipate environmental problems associated with development, we have published extensive checklists, handbooks and guides for different types of projects to help

Bank staff and planners around the world think through all the identifiable effects of development projects.

Not only is the World Bank undertaking widespread protective and corrective efforts which are being made in Third World countries, but, in addition, it is making positive efforts to enhance the environment. Although environmentalism tends to be identified with precautions in development, I believe that some of the most effective environmental efforts are those that achieve development and sustainability at the same time. Let me briefly describe three sectors where the goals of economic development and environmental enhancement coincide.

Improving systems of water supply and waste removal is one obvious area in which we can enhance the environment through development. Our lending programs in water supply and waste disposal have increased from an annual average of $300 million in the mid-1970s to more than $700 million currently. We are emphasizing economical technologies such as standpipes or handpumps for water, and septic tanks or latrines for sanitation since they cost a fifth to a tenth as much as piped systems —thus making it possible to improve the living environment of more families given the constraints of limited government budgets.

Rapid urbanization is going to continue throughout the developing world, even if vigorous rural development and population policies are adopted. We expect that the urban population of the Third World will continue to grow about twice as fast as the rural population over the next 20 years. And growing cities, if not properly managed, pose special environmental problems — especially water and waste disposal problems among the poor. So we have an active program of investment in urban infrastructure. Our purpose is not simply to help public authorities cope with current urban growth, but to introduce appropriate technologies — and to develop institutions that are better able to cope, on a continuous basis, with burgeoning city growth in the Third World.

A second sector where environment and development directly interface is energy. This area, along with agriculture and rural development, demands the highest priority among the many activities of the Bank. In addition to rapidly expanding our lending to accelerate the development of hydro and fossil fuel resources, the World Bank is also encouraging energy conservation and the development of renewable energy sources. Energy conservation has become a major theme in our dialogue with borrowing governments about their development strategies, because energy imports now represent a heavy financial burden for those developing countries which are not sufficiently self-reliant for their own growing energy requirements.

With regard to renewable energy sources, we attach high priority to forestry. While rising oil prices have captured the headlines in the United States, for almost half the world's population, energy problems take the form of a daily search for firewood with which to cook. The World Bank has embarked on a vigorous expansion of its forestry program. Our average annual lending for forestry projects over the last four years has been six times what it was in the previous four years. Two of our successful forestry projects (one in Korea and another in India) supported programs which emphasized public education to encourage tree-planting. In Burundi —where deforestation around the capital city has forced many low-income families to spend 40 percent of their incomes on firewood and fuel — the Bank has financed a promising program to introduce improved cooking stoves. Traditional stoves waste as much as 90 percent of the heat they generate, and open fires are even less efficient.

In addition to our concern about forestry and firewood, the Bank has also been exploring new ways that developing countries can increase the amount of energy they get from other renewable sources — through the production of alcohol, or, over the longer term, by taking advantage of the abundance of sunshine with which much of the developing world is blessed. The developing countries depend on traditional biological sources for a quarter of their energy requirements. And they can substitute renewable sources for another 5 to 15 percent of their energy needs before the end of this century.

Finally, in addition to water supply and waste removal and energy, a third area where the goals of economic development and environmental protection coincide is population planning. We now know that, although family planning programs help, the most effective way to moderate population growth is to alleviate poverty and to provide improved health care. Population growth rates in the Third World began to drop in the mid-1960s; birth rates have declined at least 10 percent in the world's two most populous countries, China and India. Birth rates have also declined in Indonesia, Turkey and in most of the middle-income countries of Latin America and East Asia. Quantitative analysis suggests that social and economic improvements (such as higher incomes, literacy and life expectancy) accounted for as much as 60 percent of the variation in fertility among developing countries from 1960 to 1977. The strength of family planning programs explained an additional 15 percent.

The World Bank remains committed to supporting development efforts that assist the poor to become more economically productive. We continue to invest significantly in rural development, education, health and family planning. One of the dividends of this whole range of investment — beyond the benefits to the families who are helped directly — will be a more rapid shift to a sustainable pattern of global population growth.

These are just a few examples of development which obviously contribute to long-term sustainability as well as to short-term benefits. But, as a matter of fact, we think that all development can enhance the conditions in which we live. When we finance agricultural development, measures are taken to develop existing resources and also to protect those resources for the future. When we invest in industry, given proper precautions, increased production means higher income which workers can use to improve the quality of their environment and the lives of their families. Our mandate is economic development — and development, of necessity, involves changing the natural and social environments. But with due attention to resource management, all economic development should, on balance, improve people's environment, in the broad sense of the word. This is, after all, the fundamental purpose of development.

Fairfield Osborn was a great admirer of Theodore Roosevelt. A generation before Osborn, during a formative period of economic development in the United States, President Roosevelt argued that conservation was essential to the long-term economic growth of this country. He said: "The nation behaves well if it treats its natural resources as assets which it must turn over to the next generation . . . Conservation means development as much as it does protection." Today our perspective must be global, and our concern for the environment must be even more urgent than in Roosevelt's time. His words are still true; concern for the environment is an essential ingredient of development.

The international development community must face that fact with realism and sensitivity. And I ask you who are ardent defenders of the environment to join efforts with those of us who are trying to assist the developing countries accelerate their economic growth and improve the quality of life of their societies. All of us around the world who identify themselves primarily either with the international development community or with the international conservation movement ought to be more partners than partisans. For sustainable development and wise conservation are, in the end, mutually reinforcing — and absolutely inseparable — goals.

3. British Industry and Investment Concerns On Environment in Developing Countries

Dr. Mostafa K. Tolba
Executive Director
United Nations
Environment Program
Published in 1984
Volume 4, Issue 1
The Environmentalist

Thus far, many of the environmental problems discussed have largely been geographically located in the Third World. Do these problems directly affect the United States and other industrialized nations? Four articles will help us explore this question. The first is a speech by Dr. Mostafa K. Tolba, Executive Director of the United Nations Environment Program, as he talkedwith leaders of British industry in 1983, concerning examples of global economic and ecological links and of the problems and opportunities associated with them. The speech appeared in 1984 in Volume 4, Issue 1 of *The Environmentalist.*

Dr. Tolba is a scientist who has been the head of the UN Environment Program since 1976. Having gained top levels of leadership in both science and government in his native country, he led that nation's delegation to the 1972 United Nations Conference on the Human Environment, in Stockholm, and served as deputy at UNEP to the agency's first executive director, Maurice K. Strong.

3 There is much talk now of reindustrialization, of moving to a postindustrial economy with industries like microelectronics and biotechnology coming to the fore. With the rapid growth of the new industries and the replacement of some raw materials with synthetics, I am aware we can no longer look upon the U.K. economy simply as an exporter of manufactured goods and an importer of primary products. The fact that the United Kingdom's trade figures

showed this year (1983) that for the first time since the Industrial Revolution, this country had become a net importer of finished goods, underlines the point.

However, it also must not be forgotten that now and for the foreseeable future the United Kingdom will stay in part reliant upon the developing world for food, energy and industrial raw materials. This country continues to rely for its industrial processes on a wide range of natural products such as dyes, resins, pectins, tannins, fats, waxes, pyrethrum and other natural pesticides. For the most part these come from the Third World, as do a number of foodstuffs like sugar, coffee and tea.

In turn, the United Kingdom requires markets for its output of finished goods and services. According to the Organization for Economic Development and Cooperation, 30 percent of Britain's visible exports go to the developing world.

Last year the UN Environment Program published a major scientific report on the changing condition of the world environment since 1972. It found that our knowledge of the environment in the global south is sketchy at best. But we do know enough to say that most less-developed nations are facing an unprecedented environmental crisis. Deforestation, soil erosion and desertification, misuse of water resources, and even the familiar threats in the industrialized countries of water and air pollution are undermining the human and bioproductivity of these nations.

Some aspects of poor-world environmental destruction pose obvious dangers to the economies of western nations. Take, for example, the issue of genetic diversity. The tropical forests, savannah and arid regions of the developing world are the world's genetic storehouses. (One small island in Panama contains as many plant species as are found in the whole of the British Isles.) Agriculture, pharmaceutical and even the new biotechnology industries are, to varying degrees, dependent on fresh infusions of genetic material from these wild factories. But virtually everywhere they are under threat; at a conservative estimate some 25,000 plant species are known to be threatened with extinction. Ten percent or more of all species on Earth could be extinguished over the next two decades. Many threatened species have not even been tested for their potential industrial utility, and this all relates to the species we know. Those we don't know about are many times more. Following are some examples of what the world has to lose from the loss of key habitats and species in the Third World:

• Twenty-five percent of all prescription drugs are of biological origin and come from moist tropical forests which are under severe threat.

• Alkaloid compounds, found in almost 20 percent of plant species investigated, include tumor inhibitors, anesthetics and cardiac and respiratory stimulants.

• A tree species from the Amazon basin was recently discovered to have sap which can be used directly in diesel engines.

• The entire $100 million annual Durian fruit crop of Southeast Asia depends on a single bat species for pollination. The bat species is threatened.

• The several hundred thousand square kilometers of rubber plantations in Southeast Asia were developed from 22 seedlings brought out of the Amazon nearly 100 years ago — from an area under great threat now.

•The entire U.S. soybean industry was developed from six plants brought out of Asia.

Ecological links can also be sources of conflicts

Before us looms the specter of conflict over threatened resources, over which might be termed the security of nature's supply. "Ecological linkage" is referred to in the Conservation and Development Program for the United Kingdom: "This, generally, is more uncertain and less visible than economic and political threats, but it is no less real. The ecological linkage produces direct effects upon supply, although we may feel the impact first through political instability and social strife overseas. For example, in the Third World, famine and drought have, in the last decade, started or intensified political conflicts — between Ethiopia and Somalia in the 1970s, and in India and Bangladesh. Closer to home we see not merely economic loss through ecological damage, for example in the loss of fisheries through overfishing, but we also experience the political tension that this can bring."

Following are descriptions of three regional conflicts which illustrate this dilemma:

• Fish are a hotly contested wild resource. Overfishing of national fishing grounds has led many nations to fish zones farther afield. Japan and the U.S.S.R. have competed for 40 years for the catch in the northwest Pacific. Between 1964 and 1975, 12,731 Japanese fishermen were arrested by the U.S.S.R.

• The 1975 "cod war" between Britain and Iceland was the most recent in a long-running battle over exclusive fishing zones. As Iceland extended its zone, so it entered into direct conflict with British trawlers.

• Commercial trawlers in Southeast Asia have had a disastrous effect on peasant fishermen. They have increased the total catch but brought a sharp decline in individual earnings. In Malaysia there have been violent conflicts between both types of fishermen, with boats rammed and set ablaze, and fishermen killed. [Source: EARTHSCAN, "*Environment and Conflict,*" 1984.]

With a projected doubling of withdrawals from the world's water supply between now and the year 2000, we should be alert to the danger of conflict over how nations utilize this basic resource. Of the world's major river basins, 148 are shared by two countries and 52 by two to 10 nations. We have already seen how dam-building, pollution, siltation and unsustainable withdrawals have angered nations downstream; here are additional examples:

• As many as 40 percent of the world's people live in international river basins. As demand for usable water increases, success in avoiding conflict depends on agreements regulating water use.

• Recent events in India's Punjab show dramatically what can happen when such agreement is not reached. Punjab's rivers begin in the Himalayas, flow through the Sikh Punjab area and on to Moslem Pakistan. But the Hindu Indian states south of Punjab lack water, and divert it by way of canals from Punjabi rivers.

• One of the Sikhs' demands in the recent violence which led to the storming of the Golden Temple was a demand for a greater share of river water. Punjab's wealth stems from its irrigated agriculture. It gets only 40 percent of India's agreed share of the water, and wants more.

• The prospect of climatic warming, which may well redistribute economic advantage in the production of food — the most strategic material of all — suggests that the ecological link could become the dominant connection in global interdependence in the coming century.

• The time has come when we must look upon clean air, water, top soils, wild plants, fisheries and forest cover as resources every bit as precious as, say, coal or oil or mineral ores. We have traditionally taken these self-regenerating or renewable resources for granted. These resources are renewable only if conserved, finite if not. Through a combination of avarice, irresponsibility and ignorance, we have been mining these sustainable resources. And we have been paying the price in the form of resources depletion. The world now desperately needs skilled manpower and money to move the ideas of sustainable development into meaningful action.

Solving these global problems can be good for business

New business opportunities are being opened up by the gathering interest in environmentally sound development. The UN Environment Program studies have revealed that some firms, by building resource conservation into plant development, have boosted profits, sometimes by as much as 30 percent. By vigorously applying a resource conservation and recycling program, the giant 3M Company of the United States produced total savings of $97 million from 1975 to 1981. In order to get an equivalent amount of earnings at 3M's traditional profit level would require additional sales in the order of $300 million over that period.

On a smaller scale, a distillery in Scotland, faced with an ultimatum on pollution control, decided to increase, by evaporation, the proportion of solid matter in the spent wash from its stills from two percent to 40 percent. Processing charges amounted to some £300,000 per year and capital charges to a further £200,000. On the other hand, revenue from the sales of dried wash as cattle feed amounted to over £1 million. The net profit was thus over £500,000 per year, and the pollution problems, which triggered the whole exercise, were largely eliminated in the process.

Although the science of environmental cost-benefit analysis is in its early stages and many imponderables remain, the evidence appears overwhelming that a company can use the environmental factor to boost both profits and growth. Recent estimates have put the value of the world market in pollution control and other types of environmental goods and services at some $100 billion. New firms have been set up to meet the demand, and many of the older established companies such as Exxon, Dow, Krupp, ICI, Shell and Phillips have now created new divisions to exploit this new market. And in UNEP we have also been hearing of many new initiatives on recycling wastes from countries as diverse as China, Belize, Brazil and South Korea.

4. A Priority-Ranking Strategy for Threatened Species

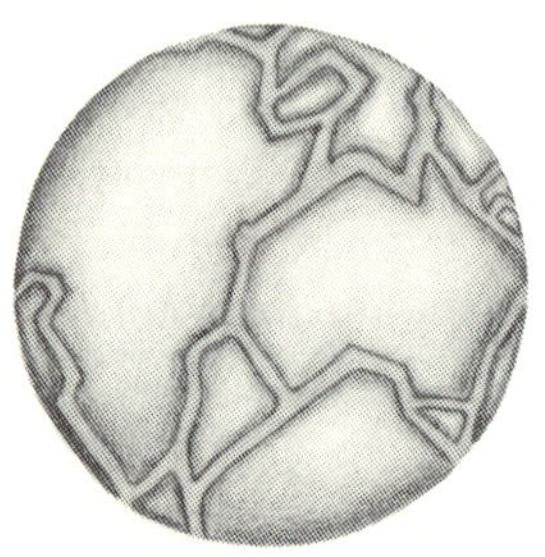

Dr. Norman Myers
Chief Editor
GAIG: An Atlas for Planetary Management
Published in 1983
Volume 3, Issue 2
The Environmentalist

Dr. Norman Myers has spent 27 years in Kenya; he is a highly respected researcher, writer and environmental advocate. Through his lively yet scholarly books — *The Sinking Ark, A Wealth of Wild Species* and *The Primary Source* - he has helped bring the seriousness and the implications of global environmental issues to the attention of decision-makers and interested citizens throughout the world. Particularly known for his research on tropical forests, he also recently served as general editor for the immense undertaking, *GAIA: An Atlas of Planetary Management*. Dr. Myers helps us understand the necessities for concern about environmental problems by suggesting that we look no farther than our kitchens — where coffee, fruit salad, chocolate, cinnamon and orange juice, for instance, are all examples of products prepared from produce grown in the tropics. He also reminds us that there is also roughly one chance in four that any pharmaceutical product we use, whether an analgesic, an antibiotic, a diuretic, a laxative, a tranquilizer or a cough drop, owes its existence, either directly or indirectly, to raw materials from tropical forests.

In this section, however, Dr. Myers is calling our attention to still another matter — the tough issue of deciding where to put our emphasis within the limited funds available to help protect threatened species. It is derived from a considerably longer article which first appeared in 1983 in Volume 3, Issue 2 of *The Environmentalist,* and which delved into several of the different types of value that threatened species have to humankind. I have chosen for inclusion here the sectors that emphasize the economic and cultural values of such species, as well as Dr. Myers' recommendations as to how the difficulty of making choices among species may best be solved.

4 The problem of disappearing species is becoming increasingly acute. Of Earth's five to 10 million species, only 1.6 million have been identified by science, and a far smaller number has been assessed for their survival prospects. Of species recognized by science to be threatened, it is generally believed that at least two or three vertebrates and two or three plants (possibly more) are becoming extinct each year. But if we consider all species on Earth, and the rate at which natural environments are being disrupted if not destroyed, it is not unrealistic to suppose that we are losing at least one species per day. By the end of the 1980s, we could be losing one species per hour. It is possible that by the end of the century, we could well lose as many as 1 million species, and a good many more within the following few decades.

At the same time, it is becoming plain that we cannot assist all species that face extinction within the foreseeable future. Conservationists have limited resources at their disposal — finance, scientific backup, etc. Even were these resources to be increased several times over, we could not hope to save more than a small proportion of all species that appear doomed to disappear: the processes of habitat disruption are too strongly under way to be halted in short order. When we allocate funds to safeguard one species, we automatically deny those funds to other species. This means that we perforce allocate our conservation resources — and thereby prioritize — to certain species in preference to others. In short, we choose in favor of some species and against others. We choose unwittingly rather than deliberately; but we choose.

Already we support only a small fraction of all species under threat. Before long we may find that we can assist a still smaller proportion of all species facing extinction. Thus a key challenge for conservationists lies in the most efficient way to allocate funds and related resources for the save-species programs.

What acould lose in terms of economic values?

Extinction of species constitutes an irreversible loss of unique natural resources, now and forever. Of the small number of species already investigated for their economic value (roughly 10 percent of all species superficially screened for any value, only 1 percent systematically screened for several values), a considerable proportion make contributions to agriculture, medicine, industry and bio-engineering. In light of experience to date, it seems a statistical certainty that the Earth's spectrum of species potentially offers many utilitarian benefits to society. In fact, species should rate amongst society's most valuable raw materials with which to meet the unknown challenges of the future. As a measure of what species already contribute, it is estimated that the American economy benefits from genetic resources to the extent of at least $30 billion per year, possibly more.

Amongst urgent questions to be asked are the following, by way of illustrative examples: In which sectors of the plant kingdom are the 80,000 known edible species to be found? Are "promising candidate" species concentrated in particular geographic localities, which will allow conservation efforts to be applied with maximum impact? Does something similar apply to genetic reservoirs that supply germ plasm for existing crops? And of plant species and strains in question, how many are already endangered, or likely to become threatened, within the next few years? Is it possible and advisable to give special attention to, for example, the 13,000 species of Leguminosae? (Many of the most important economic plants on Earth are legumes, e.g., alfalfa, beans, peas, peanuts, soybeans and acacias; and many leguminous species can support other crop plants through their ability to fix nitrogen, thus eliminating the need for farmers to apply ever more costly chemical fertilizers.)

As for medicine, wild species already supply many drugs and pharmaceutical preparations. For instance, a leading source is the organic alkaline compounds known as alkaloids. These exceptionally valuable compounds, found hitherto in almost 20 percent of plant species investigated, include tumor inhibitors, strychnine, narcotics, local anesthetics, and cardiac and respiratory stimulants. To date, only around 2 percent of the Earth's 300,000 flowering plant species have been screened for alkaloids, producing nonetheless over 1,000 forms. A key question is: Which plant groups contain greatest varieties of alkaloids? Certain families, e.g. the Rosacean and the Saxifragaceae, are known to harbor large concentrations of alkaloids useful against cancers. Does something the same hold good for other plant families? Are any of these categories of plants under greater threat than others? Which plant groups offer promise as sources of steroids, particularly for contraception and abortifacient materials? What findings are so far revealed by the investigations of the World Health Organization's Reproduction Research Unit, seeking among the plant world for a safe and effective "pill"?

A supplemental approach is to ask whether particular (geographic) areas are likely to feature high concentrations of species with probable benefit to medicine. Borneo, for example, is characterized by exceptional floral diversity, with roughly 10,000 species, many of them endemics — and at least 4 percent of them could prove valuable to medicine. The Borneo habitats of these families deserve special attention. Similarly, certain forest areas of Costa Rica have been found unusually rich in cancer-inhibiting drugs and likewise merit priority treatment.

In addition, the oceans look likely to offer many genetic resources in support of medicine. Yet the oceans, and especially those areas where the great bulk of marine life exists, i.e., coastline ecosystems, are coming under rapidly increasing threat from pollution, among other disruptions. To cite but one example, a Caribbean sponge yields a compound that looks likely to supply a breakthrough in the treatment of diseases caused by viruses, much as penicillin has achieved for diseases caused by bacteria; as a result of this widely acclaimed discovery, there is now prospect of finding cures for a wide range of viral diseases, including the common cold. Other marine species in the Caribbean might supply further compounds of unusual promise for medicine. Yet the Caribbean, as a semi-enclosed sea with growing amounts of industry along its shores — notably oil refining, plus oil extraction in several sectors of the sea itself — is highly susceptible to pollution, the latest example being the recent Mexican oil-well blowout.

Many wild species are utilized for a range of industrial products. As technology advances in a world growing short of virtually everything except shortages, industry's need for new raw materials will grow ever more rapidly. Certain plants may soon serve as sources of energy, by virtue of their capacity to produce hydrocarbons like oil instead of carbohydrates like sugars. A number of trees from the Euphorbia family, at least 20 tested to date, produced significant amounts of a milk-like sap, latex, actually an emulsion of hydrocarbons in water. Generally similar to hydrocarbons in petroleum, Euphorbia hydrocarbons are superior in that they are practically free of sulphur and contaminants found in fossil petroleum. While Euphorbia species seem to be especially suitable for "growing gasoline," some 30,000 species of plants produce latex, spanning several families. Despite the major importance of these families, next to no steps have been taken to assess the survival prospects of their component species: where are they located, what is happening to their habitats, which species offer most potential and yet are in worst shape?

A second example of industrial applications concerns lichens. A good number of Earth's 18,000 lichens, notably those that grow on tree trunks and walls, are exceptionally sensitive to traces of heavy metals and sulphur dioxide in the atmosphere (thus are excellent natural and inexpensive pollution monitors). Yet it is precisely this factor that is causing lichens to disappear in several industrialized countries. Again, a number of urgent questions need to be asked. Which categories of lichens can best serve the role of "environmental monitors?" Which lichens are most endangered? How far do these two groups overlap?

This review throws up a number of planning implications for conservation. For instance, how far can we tackle the particular case of threatened plants through cost-benefit analysis? To answer this question, let us briefly look at plants as sources of anti-cancer materials. During the past 25 years, two U.S. federal agencies — the National Cancer Institute and the Economic Botany Laboratory, both located in the environs of Washington, D.C. — have mounted a major program to track down those categories of plants that offer most promise for anti-cancer materials.

By 1980 the program had investigated some 35,000 plant species, selected from some 6,000 in general. It appears that roughly one species in 10 and one genus in four, reveals anti-cancer activity. Of the thousands of extracts with ostensible promise against cancer, only about 15 have survived numerous laboratory tests and clinical trials; and only two compounds, vincristine and vinblastine, both from the rosy periwinkle, have reached the status of "superstar drugs." Nevertheless, and as a measure of how far a single plant species can contribute to our welfare, global sales of vincristine now total around $100 million a year.

How should scientists best proceed in order to give themselves best chance of identifying key (cancer fighting) candidates among Earth's abundance of plant forms? One strategy lies with a geographic approach. Certain zones and localities ostensibly prove better bets than others. For example, tropical and subtropical climes show a higher percentage of activity, sometimes a tenfold greater proportion, than plants from temperate zones, which are higher than plants from boreal zones. The explanation for this finding could lie with the theory that the tropics feature pronounced ecological competition, the temperate zones less, the boreal zones less still.

Among tropical biomes, does any one present exceptional potential? Fortunately, there is a strongly positive response to this key question: tropical moist forests. This one biome appears to be in a class of its own as a source of anti-cancer compounds — not only because of the abundance and diversity of its plant taxa, but apparently because of the biome's evolutionary ecology which has thrown up the highest percentage of alkaloid-bearing plants, plus the highest yield of alkaloids, of all Earth's biomes. Whereas 3,000 plant species worldwide are known to possess anti-cancer properties, and at least 70 percent of them occur in the tropics, a further 70 percent occur in tropical moist forests. Thus one can speculate on the concentrated stocks of anti-cancer materials that ostensibly await exploration in this small sector of Earth's land surface, a mere 7 percent. If Latin America's forests alone contain 90,000 plant species, they could well contain 9,000 species with anti-cancer activity, some 45 of which could offer major interest to cancer research, and three of them could ultimately prove to be sources of "superstar drugs." Not surprisingly, the National Cancer Institute and the Economic Botany Laboratory believe that the widespread elimination of tropical moist forests could represent a serious setback to the anti-cancer campaign.

Curiously enough, the second most promising group of (cancer fighting) ecotypes comprise those that are at the opposite end of the climatic spectrum, viz — arid-zones. The high potential of these zones is thought to be due to the tendency for arid-zone plants, in response to environmental stress and natural "biological warfare," to produce toxins of many novel kinds — and it is these unusual compounds that can kill cancer cells. Moreover, arid-zone plants represent small and isolated families, with highly divergent, unusually distinctive biochemistries.

By virtue of their "oddball" characteristics, arid-zone plants offer outstanding promise for anti-cancer materials. For related reasons, arid-zone endemics are likewise of high interest to cancer researchers, not only because they are unavailable elsewhere but because they tend to feature unusual medicinal properties in their botanochemical makeup. So potent and diverse are the botanochemicals of arid-zone plants, that if one were to analyze 100 plants species from Egypt and 100 from Brazil, the Egyptian plants would be likely to reveal five times more anti-cancer capacity than the Brazilian plants.

For the sake of some rough cost-benefit analysis, what are the costs of the research program to track down plants with anti-cancer materials? The combined budgets for anti-cancer work on the part of the National Cancer

Institute and the Economic Botany Laboratory have amounted to only $4 million per year — a total that is to be viewed within a context of the entire anti-cancer campaign in the United States, running at just over $1 billion per year. As for the costs of cancer to American society each year, a rough estimate places the figure at $30 billion — a figure that ignores or denigrates "externalized" costs such as worker's compensation payments. While a figure of $30 billion may seem high, we should recall that cancer is the only major fatal disease whose incidence is increasing. A person born today faces a 27 percent probability of contracting cancer by the age of 85, by contrast with about 20 percent for those born in 1950.

Viewed in this manner, it would be an exceptionally cost-effective feature for American society to allocate greater expenditures to the collection and screening of anti-cancer materials from wild species of plants and animals. The market for anti-cancer drugs is expanding worldwide at around 25 percent per year, and is projected to reach $2 billion by the mid 1980s. By the same token, of course, it would be an unusually sound investment for society as a whole — not just the United States, but the entire community of nations — to assign far greater priority to safeguarding the ecosystems of those plant categories with exceptional potential for anti-cancer materials — as indeed for all plants that ostensibly offer economic benefits to humankind.

Species representing cultural and aesthetic values: Their role in protection of genetic diversity

Certain species are culturally significant, examples include the wild horses and asses, and the cameloids, as relatives of man's many beasts of burden. Certain species are beautiful or spectacular, many birds clearly qualify, as do tigers, giraffes and seals, among many others. Certain species are unusual; for example, the orangutan, the rhinos and the kangaroos. Certain species possess appeal for their "cuddly attributes"; for example, the koala bear, the giant panda and the chimpanzee.

A number of species possess symbolic significance. The bald eagle is a case in point for the United States, as is the Philippines eagle for the country after which it had just been renamed. The great whales are of special symbolic value. To the public, the whales represent creatures of unique size, intelligence and lifestyle.

The California condor represents many of the dilemmas associated with the limited funds that are available for the protection of species. With a spectacular wingspread of over three meters, the bird can soar to 3,000 meters, and live to be 35 years old. A majestic sight (on the few occasions when it reveals itself to human eye), the condor has excited a curious mystique in the public's mind. Presumably this is due not only to the condor's superspectacular appearance but to its plight as one of the most endangered bird species in the United States if not on Earth. Although the largest bird in North America, the condor has one of the smallest populations. By 1980 the species' total had declined to around 30 individuals or so, a mere eight or 10 of which are of breeding age. Despite complete protection for the birds, and the isolation of their nesting areas in high-cliff caves, the species apparently pursues a slow and inexorable march toward extinction.

How far should we go to protect a creature that is dismissed by some observers as an evolutionary "geriatric case?" Should we assume that we do not yet know how to determine when a species is inevitably declining through its own natural accord, and try to save the creature? The second option is currently being pursued, at a cost of some $25 million, spread over 40 years. (In addition, there are social costs, related to petroleum deposits in the bird's habitats that cannot be exploited; these costs can be estimated at $3 million per year.) The preservation program, one of the most expensive endeavors ever undertaken, offers, according to the U.S. Fish and Wildlife Service, only a 50/50 chance of success — and we may not know whether the investment is paying off for at least another two decades.

Despite these reservations, however, the effort may represent a sound use of conservation funds, in view of the condor's symbolic value. Suppose the bird were allowed to slip quietly into oblivion, an extinction that could be construed as part natural, part man-caused. Would the public then protest to the effect that if conservationists cannot save the condor, what can they save? In other words, does the condor perhaps possess "public opinion" value way beyond the value of its own intrinsic worth? If our best-judgment assessment tells us that this is the case, then we should go ahead with a sizable outlay of conservation funds in support of this single species, despite its doubtful prospects of survival.

The decisions about the condor are difficult in the extreme, largely because the values at stake cannot be quantified. So, at least, the argument runs. But one exercise in quantification could illuminate the issue — and to this writer's knowledge, it has not yet been undertaken. An allocation of $25 million to the condor means that a similar sum cannot be spent elsewhere. If the same monies were given over to a preservation plan for one of Hawaii's richer habitats, they might help the cause of several bird species at one time — all of them offering a better chance of success than the condor (and a Hawaiian project could protect a number of other rare and threatened species, notably plants and insects, at the same time, by contrast with the situation in the condor's habitat). Still more to the point in terms of the global heritage in species, $25 million spent on the relict montane forests of East Africa, or on the southern strip of Atlantic-coast forest in Brazil, or on certain localities in New Caledonia, would almost certainly assist several dozen threatened species — none of which, however, possesses the charismatic image of the condor.

Ways through the maze:
getting the most for the species protection dollar.

As the foregoing has made plain, certain zones are biotically richer and ecologically more diverse than others. Notable examples are tropical moist forests and coral reefs. By safeguarding a section of these biomes, conservationists can accomplish more, in terms of saving totals of species, than through safeguarding much larger zones in other biomes.

The consideration applies especially to tropical

moist forests. As a consequence of these forests' evolutionary ecology, a number of lowland areas feature high concentrations of species, many of them endemics. Conservation programs could profitably concentrate on these localities, on the grounds that they will offer a better return per conservation dollar invested than would be the case for virtually any other areas on Earth.

A similar approach can apply to other ecological "islands." The Sudd Swamp in southern Sudan, being a long-established island of moisture amid a semi-arid region, features an exceptionally rich array of species, many of them endemics. The Sudd's ecosystem is currently being disrupted by the Jonglei Canal, and entire communities of species could become extinct within the foreseeable future. Many other wetland zones can be identified, deserving varying degrees of priority treatment for conservation.

To this extent, then, conservationists can finesse the dilemmas of priority-ranked species by directing greater attention to protection of entire communities of species, and protection of entire ecosystems.

Conclusions and recommendations

Front and center, let us recognize the urgent necessity of making choices among threatened species. This is not a formidable challenge to be confronted somewhere down the road, when we have had a chance to get our act together. It is a challenge that we already cope with right now, by virtue of the funding-allocation systems that we already employ. Ever since the start of the save-species movement, we have been making choices between species. The expanded strategy proposed here amounts to no more than an extension of the past, albeit in a more methodical manner.

Those threatened species that for biological or economic or sociocultural reasons, present "the most productive opportunties" for investment of conservation resources, should clearly come top of our "shopping list" of priorities. Equally clearly, other species may not — so far as we can discern — merit such priority treatment. For lack of adequate conservation resources, and for no other reason, certain species will come pretty far down on a hierarchical ranking of priorities. Still others will be placed so far down on the list that they will effectively be consigned to a category that we designate, "We wish we could do something about them, but to our massive regret, we just do not have the means available."

The second conclusion, and associated recommendation, deals with the need to build up our information base and to refine our analytic tools so that we can better define the biological attributes, ecological attributes, etc., that pertain to the problem. Hence we should forthwith address ourselves to the task of collating and interpreting the data we need, from both the natural sciences and the social sciences, to make conservation a predictive science. Thus far, we cannot even be sure that we are asking all the right questions. But were professional researchers from both the natural sciences and the social sciences (including resource economists, environmental lawyers, systems managers and political scientists) to set about a coherent exercise of assembling the information they require, and evaluating it for priority-ranking purposes, we could, within as little as a few years, be pretty certain that we are raising all the correct questions, and we could be generating far more appropriate answers than we have achieved to date.

At the present time, under one percent of all life-science research is directed at threatened species of any kind, and only a minor fraction of this amount is directed at those generalized biological attributes that could improve our understanding of the nature of endangerment, extinction and other crucial phenomena.

A third conclusion and recommendation lies with a shift in "philosophy" of save-species conservationists. These activists could well expand their erstwhile approach by taking more explicit account of all species and their needs. The scope of save-species activities should not be restricted to those species, or communities of species, that are already recognized as threatened. Rather, it should encompass the entire array of species on Earth, with a view to determining which communities and categories of species are now in trouble, or seem likely to run into trouble within the foreseeable future. This will entail a systematic "checking procedure" right across the species spectrum, analyzing each segment of the spectrum for its resilience in the face of man-induced disruptions, traits of proneness to summary extinction and the like. As a result, no category of species will start to enter an "endangerment zone" without its plight being recognized ahead of time.

5. Converging Worlds: Environmental Events and the Free Market and Policy Development

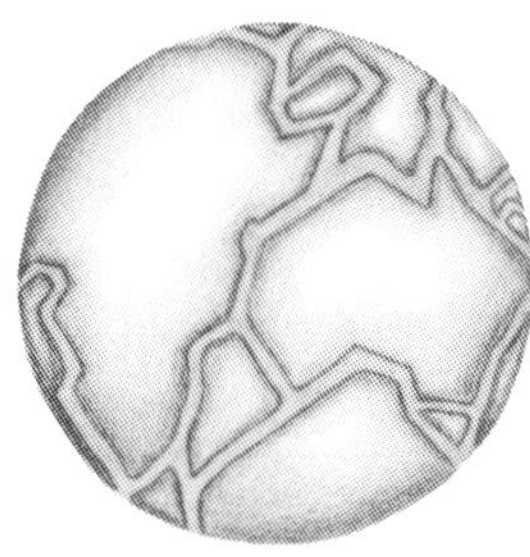

Joan Martin Nicholson Brown
Senior Liaison
Washington, D.C.
United Nations
Environmental Program
Published in 1984
Volume 4, Issue 2
The Environmentalist

Joan Martin Nicholson Brown is Senior Liaison in Washington, D.C., for the United Nations Environmental Program. Prior to her work with UNEP, Ms. Brown was director of the Office of Public Awareness for the U.S. Environmental Protection Agency. She has also served as public affairs officer for the American Petroleum Institute, and was a founder of the Bolton Institute. Ms. Brown's comments are taken from addresses she made on environmental issues to officials with the U.S. Central Intelligence Agency and the U.S. Foreign Service Institute during her tenure at the U.S./EPA during the Carter Administration. The addresses had three goals: to provide a different context by which U.S. officials might judge issues addressing the environment and environmental pollution; to offer a perspective that related this nation's foreign policy, national security and military options to environmental consequences; and to show examples as to how environmental dynamics interact with the private sector. Ms. Brown began with observations about the role of environmental affairs in the U.S. bilateral relations with Canada and Mexico in this article that was published in 1984 in Volume 4, Issue 2 of *The Environmentalist.*

5 With regard to U.S. relations with Canada, acid rain has become a problem moving into ever larger political jurisdictions and has contributed to increasing the tensions between these two countries. To demonstrate the directions such matters can take, in 1981 a rumor was widely circulated in Washington, D.C., that the reason the Canadians had focused their concern on

acid rain was that they were hopeful the U.S. Congress would respond by passing domestic legislation so greatly reducing the use of fossil fuels in the United States that the U.S. private sector would be compelled to subscribe to Canadian hydropower. No matter how contrived this rumor, the dynamics between foreign policy and the free market generated by concern over an environmental event are quickly apparent.

The issues of water quality and supply have affected relations between the United States, Canada and Mexico. U.S. relations with Mexico have long been strained as a consequence of the Colorado River arriving at the Mexican border as a mere trickle. The low quality of water, with its high saline content and pesticide residues, became such an issue that the United States built a facility at the U.S.-Mexican border to reduce the saline content. Despite this action, there is continued tension and concern as to the poor water quality and low volume.

There are, inherent in this situation, interesting international and domestic implications. If the Colorado River carried an adequate and usable water supply, what would be the possibilities for extensive agricultural development in northern Mexico? Would such a development substantially lower the flow of illegal immigrants into the United States? Might this not change the picture of the American Southwest and reduce the need for U.S. financial commitments in supporting a non-indigenous population? Could it be that the tensions over U.S.-Mexican water rights contributed to Mexico's ambivalence in responding to U.S. concerns about domestic energy supplies when U.S. supplies were threatened in the Middle East?

These vignettes not only testify to the role natural resources and environmental issues can play in foreign policy relations, but also begin to demonstrate how these issues can come full circle to affect both the public and private sectors of a nation internally. We begin to see there is much more involved in environmental issues than being for or against free enterprise or bugs and bunnies.

It might also be noted at this juncture that natural

resource/ environment issues and pollution manifest themselves in a very egalitarian fashion. Whether the pollutants are borne by biological systems or by human transport, they affect developing and developed nations in very similar ways.

If, for instance, an area of land is contaminated, a river dried up, or a desert expanded, the consequences generally are:

• the dislocation of people

• a breakdown of specific community infrastructures

• the loss of once productive lands which had provided food or supported shelter (as in the case of Love Canal)

• the loss of economic activities

In turn, these factors can and often do lead to severe pressures on indigenous political systems and leaders. Poor natural resource/environmental policies not only victimize both industrialized and non-industrialized nations, but often their effects spill over national boundaries.

For example, proper expertise was not accessible to Haiti 25 years ago when that nation's forests began to disappear because of overcutting beyond the forest's capacity to regenerate. Since that time Haiti has cut over 80 percent of the timbers from its hillsides, exposing fragile topsoil to tropical rains which quickly washed the soil into the sea. The intense sun has baked the land, which in turn has lost its porosity. This prevents the ground from absorbing and holding rainwater and restoring underground aquifers. The loss of both food and water sources has driven the Haitians from their lands to a dangerous and uncertain future in open boats on the high seas. What started out initially as a Haitian problem has become a domestic U.S. problem, as tragic environmental conditions force Haitians north to Miami as environmental refugees.

In Nepal, as in many nations of the Third World, the problem of populations expanding beyond the carrying capacity of their natural resource/environment base is forcing the poor to serve as the primary agents of pollution. The overharvesting of trees in Nepal and the terracing of marginal lands for foods, as in Haiti, is causing the erosion and destruction of the Himalayan Mountains — perceived by the world as invincible.

As the soils wash down Nepalese mountain rivers and streams into the northern plains of India, so follow the people of Nepal, adding new population stresses to the fertile but fragile plains of northern India — a reversal of the U.S.-Mexican pattern. A instabilized population too often leads to political destability, and if that population migrates to another nation, the relationship between the two can become strained or more strained depending on their previous relations.

Consider these illustrations

Among the many complex causes of conflict and wars, environmental factors play a significant and often neglected role. In 1982, Peter Thacher, Distinguished Fellow at the World Resources Institute, who was with the UN Environment Program at that time, declared: "The ultimate choice is between conservation or conflict. Trees now or tanks later." The pattern repeats itself over and over. Inequitable landholding and the conversion of agricultural land for cash (export) crops force more and more peasant farmers onto marginal lands to raise food to support their own families. These lands quickly erode, lose productivity, sometimes even reach the point of sterility. The subsistence farmers then either migrate to city shantytowns, producing destitute populations prone to revolution and violence, or emigrate abroad as "environmental refugees" — possibly increasing social tensions in their host communities.

Central America is a prime example of this process. A high percentage of refugees from Haiti, Mexico and elsewhere in the Caribbean flooding into the United States are not simply pulled by prospects of a wealthy nation; they are pushed by their inability to feed themselves back home. Large areas of Haiti are now eroded to bedrock, and Mexico's maize-yields drop every year.

Internal strife in Central America stems from the same basic factors. A 1982 draft report on El Salvador, prepared for the U.S. Agency for International Development, stated, the "fundamental causes of the present conflict are as much environmental as political, stemming from problems of resource distribution in an overcrowded land."

Similar strife is occurring for reasons of soil degradation in such countries as Ethiopia, Kenya, Poland and Peru. The political or economic system seems to be irrelevant. Nicaragua and Ethiopia are as vulnerable to destabilization from soil erosion as Haiti or Peru. [Source: EARTHSCAN, *Environment & Conflict,* published 1984.]

To look at these environmental/foreign policy relations in yet another light, in 1980 the government of Cuba became alarmed about a very large, widespread fish kill. Thousands of dead fish over a wide area of the Caribbean, but seemingly concentrated around Cuba, appeared. Cuba formally requested an investigation of the situation by the United Nations Environment Program. Cuba's fear was that this kill was the result of large-scale chemical seepage or possibly dumping. Another consideration on Cuba's part was that the kill might be evidence of an "imperialist" act, a chemical warfare tactic perpetrated by the U.S. Central Intelligence Agency against the Cuban people. In the interim, two people in the Dominican Republic had died from natural causes, but initially it was thought from eating the fish. Because the UN Environment Program, a neutral, international organization, could mobilize technical experts in the area, the cause of the fish kill could be explained. The source of the problem was two-fold — a subtle change in the currents due to intensified hurricanes, compounded by an unusual warming of the water. Had the sources of the problem not been properly identified, the opportunity for increased tensions in the already volatile Caribbean would have been heightened.

In 1979, when an offshore Mexican oil well blew, there was no Caribbean protocol in effect (as there now is) that would empower a consortium of nations to respond. The U.S. Coast Guard and U.S. oil industries have considerable expertise in managing such crises and yet their experience could not be brought to bear immediately in halting the spread of oil through the Gulf of Mexico, because of national prerogatives in territorial waters. (Texas subsequently sued Mexico as oil debris washed up on the Texas coast.) The gulf was held captive to an environmental crisis that only the good fortunes of winds, tides and waves could resolve. Ultimately, the Mexican government did request the assistance of the U.S. Coast Guard and the Environmental Protection Agency. Because of favorable weather conditions and international cooperation, tourism and fishing were granted a reprieve from short-term catastrophic damage, but the verdict is still out on long-term effects.

The practice of foreign oil ships purging their bilge in the Caribbean can quickly affect the level of island tourism, if oil and tar balls wash up on island beaches. A downturn in tourism can mean a downturn in an island's GNP quite quickly. It may do more damage than foreign aid can offset. (And may affect repayment of loans to international lending institutions and/or commercial banks in the United States and elsewhere.)

The climate of diplomatic relations often feels the effects of environmental issues. The attention focused on deforestation in Brazil by U.S. non-governmental organizations and the U.S. media introduced a new tension into U.S.-Brazilian relations. Brazilians perceived the world's limelight on the management of their forests as an intrusion on their sovereign rights. Whether such concern is warranted or whether the intrusion is justified is not the issue, but it points out, again, that foreign policy conditions are not immune from environmental issues, for the realm of concerns is even wider.

In 1979, in another part of the world, Sierra Leone received an offer of funds by a U.S.-based chemical company if Sierra Leone would facilitate disposal of a hazardous waste known to be a proven human carcinogen. The chemical company had decided it would be cheaper to export their waste than to comply with the U.S. Resource Conservation and Recovery Act regulations. Through a fluke, the offer was discovered by a Sierra Leone student studying in the United States, and made public. With enormous poverty in Africa, the desperate need for hospitals and schools, roads and jobs, one could understand Sierra Leone considering the offer. However, when the U.S. State Department, learning of this proposition, informed the government of Sierra Leone of the potential problems, the transaction was halted. This illustrates how quickly decisions traditionally believed to reside solely in the private sector fall into the arena of public policy.

In this scenario some interesting dynamics can be examined. What if the plan had gone undiscovered and the shipment was made? What if five years later, or 40 years later, the United States, for strategic reasons, required the use of a military base or port in Sierra Leone? At that time a health epidemic might break out which could be traced to the illegal dumping of this hazardous waste (one can too well recall the political fallout of the Love Canal). The United States hope for cooperation with Sierra Leone would be seriously jeopardized.

In another scenario, what if Sierra Leone uncovered a large supply of a strategic resource, but refused to sell to the United States as a consequence of domestic pressures stemming from problems with the dumped hazardous wastes?

As a third scenario, what if other U.S. corporations wished to produce or market in Sierra Leone? Might there be a stigma against other U.S. private-sector initiatives as a consequence of one Love Canal-type disaster?

These scenarios, once again, portray how a nation's foreign policy, its options, military as well as economic, can in the long term be affected by environmental events.

Signs of hope

The Mediterranean, the long-time showcase for an enclosed ocean, had moved from being the shared resource of nations for fishing and tourism to serving as a common cesspool. Although the Mediterranean was serving a useful waste-disposal function for some industrial activity around the area, subsequent increasing pollution was destroying other industries, specifically fishing and tourism. As cattle ranchers and coal miners have come to conflict in the United States over land use, the nations of

North Africa and Europe came to conflict over the Mediterranean. Before the Camp David accords and in spite of historic animosities between Turkey and Greece, Spain and Morocco, Algeria and France, and others, traditional foreign policy positions were set aside to halt the pollution through the Regional Seas Program for the Mediterranean. The degradation of that historic water body came to be viewed as a matter affecting each nation's economic security.

Thus we can see that environmental problems are changing traditional foreign policy perceptions and can, if properly responded to, open up new options for nation states in their relations. But, if not properly responded to, domestic and international problems which stem from environmental/natural resource issues can strain both national and international institutions and their capacity to respond. As a nation's land, air and water systems become bankrupt, the resource base disappears and can lead toward the economic bankruptcy of a nation.

No one is better able to understand the consequences of bankruptcy than those representing free-market systems, for political and economic instability is the antithesis of what must exist for free-market institutions to flourish. How people affect land, air and water systems is indeed the bottom line for how national and international trade and the free market system will operate.

Current events are littered with domestic and international issues stemming from economic, political, social and environmental/natural resources matters, connected in ways which too often we refuse to acknowledge or understand. It is imperative that we come to understand and consider these connections.

If we are to act on this understanding, we must increase our attention to those long-term environmental improvements which help stabilize nations and their economies. While we cannot and should not neglect the world's ecological refugees, who are requiring more and more immediate foreign aid to sustain their very lives, we must realize that in the long run, environmental and economic development projects are equally necessary and must evolve simultaneously.

6. One American Community's Link to the Third World

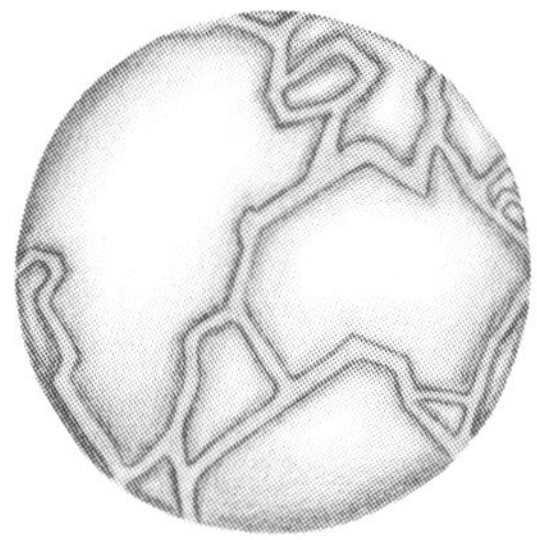

Published in the *World Bank News*, Vol. IV, No.7. .

This final selection in our exploration of the nature of global environmental issues comes from the World Bank, which recently held an experiment to prove that American newspapers can, by looking at their own communities, see how the communities are related economically to the Third World. The following story appeared in the *Hattiesburg American* (Mississippi) during the experiment. In the article, the global links we have been exploring turn away from top-level international politics and multinational corporations to focus on the relationships that have arisen between individual American farmers and Third World countries. We even get some insights into the time when the United States itself was a developing country.

6 Under the broad oak trees on Joe Morgan's farm, south of Hattiesburg, where freshly harvested fields of soybeans spread out to a rolling sea of pine forests, the Third World seems as irrelevant as a mule-drawn plow. But poor countries of Latin America, Asia, Africa and the Middle East are as important to Morgan as the computers and sophisticated farm equipment he uses to make a living. Although his crop is soybeans, he has never tasted soy sauce, a soybean derivative as important to East Asians as ketchup is to Americans — and he's never visited Mexico or Nigeria, where soybeans are used to make feed for chickens.

Morgan hauls virtually all his crop to the Pascagoula waterfront where from the storage bins run by the French-owned Louis Dreyfus

Corp., one of the world's four largest grain merchants, his produce is loaded on vessels bound for every region of the world. His Third World ties are so critical and complex that it is not enough for him to check with agricultural brokers daily about the world price of soybeans or of winter wheat, which he also trucks to Pascagoula. He must anticipate international economic and political trends such as whether Brazil will enter the soybean export market, as it has, or if the Soviets will invade a Third World country and prompt the American government to impose a grain embargo. Morgan calls his ties with the Third World, "complicated."

So it is for the rest of America, which lives in the same interdependent world as Joe Morgan. He is only one of 400 farmers, most from south Mississippi, who bring their soybeans and wheat to Pascagoula, says Ron Cox, the buyer at the Dreyfus export elevator there. Morgan and his fellow farmers are, in turn, important to the health of the local economy. The Mississippi Cooperative Extension Service calculates that each dollar he loses in export sales takes $2.50 out of the local economy. County agent Malcolm Broome estimates that, overall, half of Forrest County's $20 million agriculture producion is exported. And agriculture is only one of the strands that tie this South Mississippi area tightly to the Third World. These ties — and political, social and economic ties as well — promise to multiply in the future.

Mississippi, like America generally, was, in its infancy, as underdeveloped as many Third World countries are today. In the last century the state's economic strength lay in exports of raw materials, chiefly cotton, to economically advanced nations in Europe. In the search for capital to underwrite its own development, the state turned to these same nations for loans. British investors helped finance the railroad lines that put Hattiesburg on the map. With the arrival of economic bad times, not unlike those that besiege the world today, Mississippi's legislators passed a constituional amendment defaulting on foreign obligations in the 1870s.

Today, in contrast, Mississippi's largest bank, Deposit Guaranty, has $32 million in outstanding foreign loans, much of that to financially troubled nations in Latin America. E.B. Robinson, Jr. chairman of the board of Deposit Guaranty, says these loans are solid and will not end in default. But international economic problems clearly overshadow Joe Morgan's prosperity. The once-robust export market for United States soybeans has decreased by about 20 percent in the last two years, Dreyfus officials say. Two of the key reasons are weak foreign economies, which are unable to import, and the relatively strong U.S. dollar, which makes American goods more costly abroad.

The international oil crisis has added to Morgan's woes, causing the prices for diesel fuel to quadruple during the past decade. He uses 28,000 gallons a year to run his equipment. Banking problems, domestic and foreign, have pushed his interest rate to 15 percent. Meanwhile, Morgan needs more than $250,000 a year to finance the planting and harvesting of his crops. Floating interest rates on his loans for land purchases have increased by nearly 50 percent in the last decade. A critical factor for Joe Morgan's future is the Third World, according to Joe Petrowski, who follows international prices for Louis Dreyfus in the United States.

South Mississippi farmers find it difficult to compete domestically with farmers in the Midwest, where the soil is richer, but they do have an edge in exporting because of their proximity to export terminals. While European imports are not likely to increase, the Third World will buy more if it can pull itself out of the current spate of economic woes, analysts say. Meanwhile, says Morgan, referring to the complex world in which he lives, "every decision you make these days has to be good."

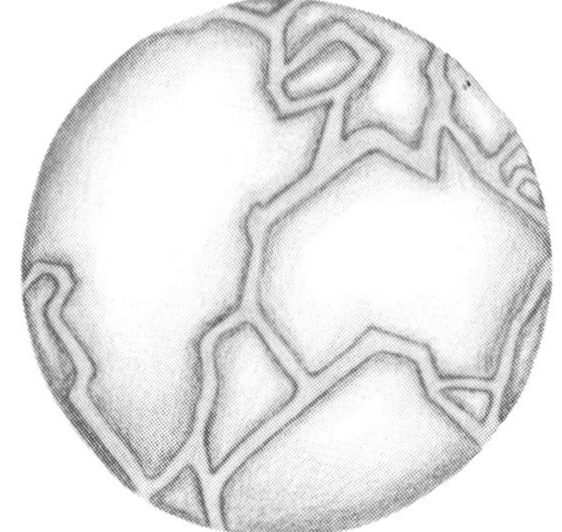

PART TWO

"Much of our disorientation today arises from the fact that the controlling assumptions of our society have now been fundamentally altered. Against the presumptions of the past we are painfully finding that resources are not infinite, that markets will not necessarily remain open or raw materials available on terms favorable to industry, that technology is not always benign, and that the earth, sea and biosphere cannot absorb the unlimited outpourings of economic growth . . . If maturity is the recognition of one's limitations, so also is societal maturity perhaps a recognition of our limitations. We must do some hard thinking about the future."

"Interdependence in a Changing World"
Colorado Governor Richard D. Lamm
An Address to the U.S. Council on Foundations
April 25, 1984

Key U.S. Pieces of the New Global Puzzle

This second section of the book looks in greater detail at the ways that several of the global environmental problems are manifesting themselves in the United States.

The four papers included give a sense of the frustrations associated with any search for solutions. Particularly within the United States, threats to public health and safety and/or natural resources are usually the sparks that ignite citizen demand for action — fast action. But most of these problems represent new types of challenge to which our system is not well equipped to respond. Being a nation of immense natural resources and space; we have tended to assume that the air, water and soil can absorb the impacts of our societies and will continue to be available, uninterrupted and in un-degraded states, to meet our needs. We have not anticipated many of the emerging problems of waste disposal and pollution cleanup, or the needs for pollution prevention and resource conservation. The technical and scientific background information, and even the legal formats for actions that can accomplish the most with the least negative impact, are seldom available. We are having to learn through trial and error: A method that is expensive and is creating still other frictions and concerns as the uneven impacts of attempted solutions — on the vitality of various economic sectors, on personal freedom and lifestyle, and on equity among parts of the country and elements of our society — ripple out across the nation.

The papers that appear in this section are examples of the processes through which our communities, states and nation are going as we struggle to understand and adjust to these new environmental realities. For readers interested in more detail on any of the programs described, the authors and appropriate agencies are listed at the back of the book.

The four papers explore the following:

- Agricultural lands preservation: a key aspect of being able to meet existing and growing global needs that can feed the world's growing population and also improve the quantity and quality of food for the world's existing poor.
- Dealing with the exposure of people to toxic chemicals: a problem whose dimensions and implications need to be clarified at the same time that we implement, then refine, improved management of these substances, which have become an important part of our agricultural, manufacturing and everyday living processes.
- Addressing air and water pollution: here examined as types of problems that can cross national as well as state and local governmental borders, creating political as well as scientific and technical challenges.
- Assessing progress: looking at what has happened thus far, within the United States, in dealing with the range of environmental problems whose importance is being recognized.

The papers were selected for the thorough way in which they describe the many interacting factors that are present within individual environmental problems. Collectively the papers serve as another reminder of how interconnected environmental issues are — with each other, and with other societal concerns. While preservation of the U.S. ability to produce food is vital to this nation's well-being and its strength in the world economy, poor agricultural practice can be sources of toxic contamination, as well as causes of polluted air and water. Likewise, unsafe disposal of toxic substances can render prime agricultural lands, as well as water and air supplies, useless for human purposes, release of polluted air, or of polluted or simply uncontrolled flows of water, can likewise affect the productivity of agricultural and forest lands, and public health and safety. These issues demonstrate why our success in finding solutions is so important to all who live within the United States. Our experiences, as we try to deal more effectively with these complex interacting processes within our own nation and political system, helps us understand the need for both sensitivity and persistence as the world attempts to address these same factors on a global scale.

1. Saving Agricultural Land: Environmental Issue of the 1980s

R. Neil Sampson
Vice President
American Forestry Assn.
Published in 1982
Volume 2, Issue 4 of
The Environmentalist

1 The major public effort in the United States today is focused on getting the economy back on track and reducing the size and expense of the federal government, but there are warning signs of far more basic problems on the horizon. America's agriculture is in dire straits, and unless this basic industry can be guided back to a more sustainable course, the entire nation is in jeopardy. At risk is not only the financial security of American farmers but also the inherent productivity of the lands upon which agriculture depends. Some observers have called this "the most important item on the list of unfinished business for the environmental movement in America." I would also classify it as a major item of unfinished business on the nation's economic agenda. That makes it a vastly different issue than many environmentalists have faced in the past, and those differences affect the manner in which the issue must be addressed.

At stake in this issue is the basic ability to survive, both as a people and as a nation. Certainly, using farmland raises questions like the loss of options for future generations, increased air and water pollution which contributes to loss of environmental quality and health today, destruction of scenic rural landscapes, and the loss of rural social and cultural values. These are important environmental and social issues, similar in nature — and often similar in cause-and-effect — to the issues of recent decades.

But there is also the urgency of keeping the United States food system, and therefore, the United States, in business. This is a question that was simply not heard in the public forums of the 1970s. It is an issue with critical political implications, not limited to farmers, environmentalists or any other interest group. It concerns us all, and affects our daily lives more directly than many of the open space, wilderness, wild river, or air- and water-quality issues of the past decade.

At the same time, the farmland issue bears some remarkable similarities to other natural resource issues of recent years. We face an era marked by land limits in place of surpluses, and that is the same reversal that characterized the petroleum situation in the United States in the 70s. It may be that we can gain some insight into the farmland issues of the 1980s by reviewing some of the economic and political responses to that energy situation.

It is scarcely necessary to argue that a healthy, productive, profitable agriculture is essential to the public interest of any nation. In the United States, significant foreign policy and trade issues are involved in addition to the normal domestic food interests. U.S. farm exports were in the range of $40 billion in 1981, despite a worldwide recession. Thus, farm commodities remained a major bright spot in an otherwise gloomy economic picture. But while those export figures brought smiles to national statisticians, they caused little joy on the farm. United States Department of Agriculture economists calculate that it cost an average of $5.20 to produce a bushel of wheat in 1981, but that wheat was sold on the international market for $3.80, and the domestic price was established at the same level.

The result, predictably, was a deepening financial crisis on America's farms. U.S. farmers were in debt $175 billion at the start of 1981, and slipped another $20 billion during the year. Good growing weather in 1981 resulted in too much product to be absorbed in a world economy buffeted by recession, and national policy provided no relief for farmers. The outlook for 1982 is even worse, with net profits predicted in the $10-15 billion range, the lowest real purchasing power on America's farms since the Great Depression of the 1930s. [By the end of 1985 the current farm debt had risen to $215 billion].

The connection between this grim financial picture and the survival of the nation's farmlands is devastatingly simple and direct. Farmers are in business, and when any business is losing money, it has but two places to turn: depreciation of its capital assets (its lands) or extension of its credit line. Farmers have been doing both, and soil-damage rates have been accelerating, and for the same reasons, as debt levels have skyrocketed.

Before we turn to some of the basic causes of these problems, it may be helpful to briefly detail the current resource pressures on farmlands in the United States.

Soil erosion

Despite a significant investment in conservation practices by the federal government and by private landowners, erosion from agricultural lands continues at a massive rate. Nearly 4 billion tons of topsoil are moved each year by sheet and rill erosion, with half the loss occurring on cropland. Adding the losses caused by wind, gully and streambank erosion brings the national total to nearly 6 billion tons of topsoil moved each year.

No one can fully conceptualize what 6 billion tons of topsoil actually means, of course. One interpretation, based on the general concept of tolerable soil loss, is that on 12 percent of the nation's croplands, and 17 percent of its rangelands, we face the task of slowing soil erosion or writing those acres out of the productive inventory within a few short decades.

In the past, the adverse effects of soil erosion on productivity have been masked. New and more productive crop varieties coupled with the heavy use of inorganic fertilizers, better control of pests and crop diseases, and improved tillage and planting methods resulted in yield increases despite topsoil loss. While those technological increases helped hide the effects of soil erosion, at least temporarily, they did not cancel out the real losses. If farmers continue to let topsoil slip away at the rates found in 1977, they are doomed to a future of spending more and more trying to coax less and less from a dying land.

Although it is difficult to predict with any certainty,

the data at hand suggest that over the next 50 years the loss of productivity due to erosion on United States cropland will be equivalent to the loss of between 25 and 62 million acres. To get some idea of the magnitude of that loss, 25 million acres could produce 50-75 million metric tons of grain, or half of the total exported from the United States in 1980. The loss of 62 million acres could represent the loss of virtually all of 1980's exportable surplus. So these estimates are significant to America's productive future.

Farmland conversion

Every hour in the years between 1967 and 1977 saw about 320 acres of U.S. agricultural land converted into non-farm uses. Not all of that land was cropland, but about 100 acres were, and another 100-120 acres had the physical capability to be used for cropland. Losing 220 acres of existing and potential cropland an hour means over 5,000 acres every day — the equivalent of losing 23 average farms. It means an area the size of the average Midwestern county every 2 1/2 months, and the equivalent of a whole state by the year 2000. It means paving over a one-mile-wide strip from New York to California every year.

The loss can be calculated in economic terms as well. Each day America's potential to produce corn was reduced by some 5,800 tons, worth over $620,000 when corn is $3 a bushel. Each year we lost the capacity to produce about 2 million tons of corn, worth some $220 million at the same market price. And each year's losses must be added to the previous year's losses, and the next year's losses will add to both, until the total economic impact from wasteful land-use practices adds up quickly to an appalling economic toll.

Urban uses take the best land. Between 1967 and 1975, about 39 percent of all the land converted to urban use was prime farmland. Since prime farmland makes up only about 25 percent of the nation's non-federal lands, it is clear that urban development tends to concentrate on the best lands.

The rate of land-use changes affecting agricultural

lands speeded up dramatically in the 1970s. By 1977 there were 90 million acres of urban and built-up areas larger than 10 acres, up from 61 million acres in 1967 and 51 million acres in 1958.

There is not much hard evidence to suggest that those trends will slow down, but there are nagging signs that Americans, while they may not think small is better, are beginning to believe that it is all they can afford. The current economic situation may not be a short-term belt tightening, but the start of a very different society in which we build fewer roads and airports, are less wasteful of land, energy and community values in settlement patterns, and settle for smaller living spaces using less land per person.

Such trends would slow land conversion dramatically, but even they would not alter the fact that there will be 11 million more Americans reaching the prime home buying age of 30 in the 1980s than did so in the 1970s. The number of new families will increase 25 percent in the next decade. When added to the number of families occupying substandard housing and the number of housing units destroyed each year, the net result could be the need for 23 million new housing units in the 1980s. That means a significant continuation of the pressures on American farmlands, even at lower per-capita land-use rates.

Soil quality

In addition to the losses being inflicted on our farm productivity by soil erosion and farmland conversion, there are other resource concerns as well.

Soil organic-matter levels appear to be dropping in the humid cropping zones such as the Corn Belt, and soil compaction appears to be increasingly troublesome, although there is little data upon which to base solid conclusions. The 1977 National Resource Inventories done by the Soil Conservation Service did not measure organic matter levels, and there is little benchmark data against which to compare such measurements if they were done.

In U.S. arid lands, desertification is affecting somewhere in the range of 225 million acres, ranging from the southern tip of California, east through southern Nevada, Arizona and New Mexico, then almost all the western half of Texas, and northward through the Oklahoma panhandle into southeastern Kansas. Increasing salinity is impairing the productivity of some irrigated cropland in the 11 western states. Salt build-up in the soil is difficult to reverse. It can be done, but it will increase costs and result in that land becoming less economically competitive, particularly for lower-value crops.

In some parts of the nation acid rain is affecting crop growth. Data assembled has shown that acid rain affects part or all of the northeastern United States, in addition to large areas of the West around Los Angeles, Oregon's Willamette Valley, Tucson and Grand Forks. The damage potential, especially in humid areas, is massive, and the problem is spreading.

The water situation

Agriculture is by far the largest single consumer of water in the United States, accounting for 83 percent of total water use, some 89 billion gallons of water each day. Irrigated agriculture produced 27 percent of the value of farm crops harvested in 1980, on only 12 percent of the harvested acres. Most of that irrigation takes place in the 17 western states, which harbor five out of every six irrigated acres. There is increasing irrigation throughout the rest of the nation, however. By 1977 every state except New Hampshire and Rhode Island reported some irrigated acreage.

The Water Resources Council has estimated that water consumption will increase about 27 percent between 1975 and 2000. Most of this added demand will come from growth in manufacturing and mineral industries, steam-electric generation, and agriculture. In regions where water supplies are limited, that means increasingly intense competition for the available surface water and groundwater.

Our ability to increase water supplies has about run its course. The Water Resources Council estimated that, by the year 2000, there will be severely inadequate surface water supplies in 17 subregions, located mainly in the Midwest and Southwest. During dry weather, more sub-regions, including some in the East, will also be faced with a water shortage.

In recent years groundwater has been available to fill in the gaps where surface supplies were overtaxed, but this source is dwindling as well. Spurred by improvements in pumps and reductions in energy costs, groundwater use increased to the point where it was providing 39 percent of all irrigation withdrawals in 1975. But much of that water is being pumped from pools where the water supply is replenished more slowly than the current rate of withdrawal. In 1978 the U.S. Water Resources Council estimated that this "mining" of groundwater was depleting the nation's water supplies at the rate of 21 billion gallons a day. Past increases in groundwater available for agriculture seem certain to be replaced in the future with substantial decreases.

In total, it is doubtful if irrigation will make the contribution to the expansion of agricultural yields that it has in the past 30 or 40 years. Rising energy prices are expected to double the costs of irrigation with a standard center-pivot system between 1980 and 2000. Efforts to improve system efficiencies may drop that cost somewhat, but there will still be an economic disadvantage facing irrigated agriculture in the future.

Total demands on the land

Each of the resource pressures, viewed singly, portray but part of the pressures on farmland. Adding them up gives a more complete picture, but even then the situation must be viewed in the light of the total land resource available to American farmers.

In 1977 U.S. farmers used 414 million acres of cropland, including all the land that was in summer fallow, rotation hay and pasture, orchards, vineyards and similar uses. There were estimated to be 127 million more acres of land that could be converted to crops on an economic basis. That makes a total of about 540 million acres of land that could be counted as our total (U.S.) cropland resource pool. From that pool we are steadily losing between 1 and 3 million acres each year as farmland is converted to non-farm uses, another 1 to 3 million acre-equivalent each year to soil erosion, and an unknown amount to such factors as soil compaction, desertification, salinity, alkalinity, water/ logging and the loss of irrigation water supplies.

Based on the rates of loss experienced in the decade between 1967 and 1977, that 540 million acres could shrink to somewhere in the range of 500 million acres or so by the year 2000. If average crop yields continue to turn upwards at the rate of about 1 percent per year, producing the basic food and fiber that we might need in 2000 is expected to take somewhere in the range of 450 million acres, providing that farm exports do not grow significantly between now and then and there is no additional land needed for energy or industrial crops. If exports continue to grow as they have in the last few years, or if energy and industrial crops take the 20-50 million acres that it looks like they will need, it is easy to see that all the land available will be used.

Future yield increases, if we can realize them, could help us meet vastly larger demands. If yields rise by 2 percent a year, we could meet USDA's demand projections in 2000 with around 400 million acres of cropland, or we could sell a great deal more crop overseas. That would just about eliminate any need for concerns about our farmland supplies, at least for this century. If we don't get those yield increases, however, we are in trouble. If average crop yields in 2000 are no higher than today, the land demand will be in the range of 570 million acres, far more than could be economically developed unless the price of food more than doubles.

Which one will occur? That, of course, is anybody's guess. The U.S. Department of Agriculture has estimated that yields may continue to grow at about the 1 percent rate, provided that federal spending for research grows about 3 percent per year in terms of real dollars. There doesn't seem to be much chance of that, at least in the near future, but then there is not any certainty that research is the real problem today at any rate. Our main problem may be economic depression. If the dollars needed to buy the fertilizer, seed and pesticide for today's agriculture were to become seriously limited, our average yield would not just stop rising, it would plummet like a stone.

The fact that today's predictions from USDA and the agricultural economists indicate reasonably good food security in the United States until the year 2000 (barring unforeseen difficulties) is no particular cause for rejoicing. The corollary is that if any of the assumptions prove to have been overly optimistic or situations change at all, we could, in the future, regret not having taken available conservation options today. If the USDA predictions are correct, shortages are likely some time shortly after 2000 unless major unforeseen technological breakthroughs occur. Ask someone who is 20 how they feel about that kind of future.

The real point is that wasting good farmland is poor business, both for current well-being and future security. It cannot be defended on any grounds — legal, moral or economic. Americans are wasting the greatest body of prime farmland on the face of the Earth, and that is both senseless and suicidal.

Interpreting resource trends: An uncertain science

We face a future that will test our skills and ingenuity; one that will be far different from the past. The age of resource abundance in America is rapidly drawing to a close. We must start to think about land, and the technologies that use and manage it, in an entirely different manner than we did only a few years ago. We must, in short, learn how to think about limits.

It would be a mistake, however, to be overly simplistic in any evaluation as to where the current land trends are leading. Many previous analysts have been proven wrong by making predictions that were too apocalyptic. We need to look more closely at the very nature of resource constraints, and how they manifest themselves in today's economic and political system. The energy issue in the 1970s offers a fresh example that can give insight into what may emerge in relationship to land and food.

It was clear to most people in 1970 that the spectacular increases in per capita energy demand and usage in the United States could not be accommodated for very much longer, but that idea was by no means universally accepted. Observers with outstanding academic and professional credentials were taking widely different attitudes about the potential problems facing the American public.

Leading the parade to tell us that we faced no problems were the resource economists. In 1972 staff members at Resources for the Future did an intensive study for the White House entitled "Population, Resources, and the Environment." In it, they assured us that "even without assuming any breakthroughs, world fossil fuel reserves (including potential as well as proven) appear adequate for at least the next half century." The United States had adequate reserves, said Resources for the Future, even without taking shale oil reserves into account, to make up for any disruption of foreign supplies. United States imports, to be sure, were rising as a percentage of total consumption, but that "should not necessarily be taken as a sign of increasing long run vulnerability."

That report would not have made popular reading for the American motoring public waiting in the gasoline lines of 1974. To be sure, the RFF authors may still argue that "long run" means more than a decade, but the fact remains that wellhead prices for oil increased tenfold, and gasoline prices at the retail pump tripled, in but a few short years. The decade since 1972 has seen supplies and prices bounce up and down, but the general direction has been toward sharply higher prices, with few signs that petroleum will ever return to original price levels.

But who could have looked at events of the previous two decades and predicted such an outcome? Petroleum had been steadily dropping in relative cost since World War II. American automobiles had been getting steadily larger, heavier, more powerful and more luxurious. Where would one have found evidence in 1970 to predict that, in 1982, we would be paying $8,000 for a small car which we would drive far less than before because, even with greatly increased fuel efficiency, the cost of owning and operating a vehicle had tripled? Certainly, the American car manufacturers missed the point. "Small cars mean small profits," Henry Ford II was quoted as saying. Today "expensive cars mean no business," would be a more accurate lament in that beleaguered industry, one which directly or indirectly is said to be responsible for one out of every six jobs in America.

Why had such respected economists as those at Resources for the Future failed to warn us adequately of such a turn of events? Dennis Meadows, famed for his work with the Club of Rome 1981 study *The Limits to Growth,* has offered five reasons why this might have happened. As we review his critique, it is obvious that some of the same situations apply to the evaluation of the farmland and food situation today.

First, Meadows argues, the RFF researchers failed to understand (as, he might have added, did many others) the drastic impacts that even modest rates of exponential growth can have on a resource stock. He points out that a resource supply that could last 400 years if the current rate of consumption held steady, would last but 75 years if the rate of consumption were growing at a rate of 3 percent annually.

Many people will not apply this notion to the farmland supply, arguing that land is not a "fund" resource, in the same sense as petroleum or mineral ore. There is no need to "use up" the land in the process of producing food. Soil productivity is a "flow" resource that we can use this year, and still have next year. History bears this out. There are lands in many parts of the world that have produced food for millennia, and still seem perfectly capable of continuing to do so.

But that is not the whole story. In America today one-third of our cropland is being allowed to erode at rates faster than topsoil can be replaced. On much of the remaining cropland, the topsoil is losing essential nutrients, structure, water-holding capacity or organic matter in the process of crop production. That land is, in fact, being "mined." The land damage estimates that we have reviewed are clear evidence that American agriculture is, for the most part, using its land as a "fund" resource rather than as a "flow" resource. The implications of that, when compared to the total world-population growth rates and the suitable supply of arable land, are grim. Using the land as a non-renewable resource assures an eventual collision between food supply and demand, and even modest rates of exponential growth in population hasten that confrontation.

A second factor involved is the degree to which population and economic growth rates accelerate competing demands on the land. The 1975 and 1977 surveys conducted by the Soil Conservation Service came as a serious shock to many who had been watching the urban and industrial uses take only a small, seemingly insignificant part of the landscape. The revelation that conversion rates had tripled in a decade were met with complete disbelief in many quarters, and are still the object of some contention. But disagreements over the veracity of the data may be largely irrelevant. What may be critical is not whether or not there were 3 million acres a year added to the urban and industrial land uses, or whether we should count 1 or 2 million as being the most accurate estimate of the loss of significant agricultural production potential.

What is critical is the reduction in commercial farming opportunity in whole communities or regions as a result not only of urban growth but of urban growth patterns. It is not unusual to find that five to 20 acres have been compromised for each acre converted in a typical "buckshot" pattern.

Urban developments, strip mines, highways and reservoirs change water regimes, disrupt rural communities and overload rural service systems. Each of these factors, while perhaps not affecting the farmland supply, diminishes farmland productivity — in significant ways.

It is little wonder that farmers feel threatened by rapid population growth and land-use change. They understand — perhaps only intuitively — the added costs and burdens that often accompany this growth, even when economic analysts can't find a trace of it.

A third area where forecasters of resource problems can make horrible miscalculations involves the price elasticity of supply. If prices rise because one resource is limited, another source of supply will become economic and fill the need, theory tells us. But how easily will that occur?

Remember shale oil? When world oil prices were $4 a barrel in the early 1970s, we were told that oil from shale would be economic when prices hit $8. Later in the decade, when oil was $11 a barrel, we were told that $23 was the magic number. Today, with OPEC oil in the $30 range, we still aren't in the shale oil business.

That outcome is easily explained. Producing oil from shale is an energy-intensive technology. As the basic price of energy — reflected by the price of oil — rises, it drives up shale oil production costs apace. Oil from shale may be like the carrot dangling from the stick in front of the donkey's nose — always tantalizingly just out of reach, but leading us steadily forward in a vain effort.

We have the same application of resource supply theory on the farmland issues. "What does it matter," some say, "if we lose a million acres to urban uses, or if 10 tons of topsoil leave an acre of cropland. We'll just plow another acre or pour on additional nutrients to make up for the loss." The problem is that the rising price of energy is raising agricultural costs as well, and clearing a new acre may not be nearly as economically feasible as it once was. In fact, it may be a great deal more expensive than keeping the former acre in crops and subsidizing urban development onto non-cropland. Fertilizer costs have doubled in the past decade and, if current plans to deregulate natural gas ever go into effect, the cost of nitrogen fertilizer could redouble in very short order. Any evaluation of the ease with which other resources can be substituted for the inherent productivity of existing farmland must take the rising costs of production created by the energy situation and continuing resource deterioration into account. In addition, we find that the same cost:price squeeze that hastens the abandonment of cropland makes its replacement less feasible. Farmers who find it unprofitable to produce crops today on soils that often have two-to-three generations of development and improvement costs sunk into them will find it even less profitable to make these investments anew on previously uncropped land.

The fourth factor involved in estimating the elasticity of the land supply in relation to rising prices is based more on qualitative factors than quantitative, and thus is harder to evaluate. Tremendous land-use shifts have marked U.S. agriculture in the last three decades. In general, commercial farmers have fled the northeast and Appalachian regions to develop new croplands in the Mississippi delta, the plains states and the West. Those new lands were usually in large, flatter chunks that were better adapted to current farming methods and machinery. Irrigation offered the assurance of adequate moisture supply, taking a large part of the risk out of high-input, intensive techniques. But the supply of unused good land, in large, flat parcels, in good climates with adequate water resources, is about used up. What was done in the past can't be repeated. Substitution of land resources is not just an economic process. It has both economic and environmental limits.

Much of the recent cropland development has been spurred by economic incentives created by quirks in tax law. One economist has calculated that, in the Nebraska Sand Hills, the tax code rewards the conversion of rangeland to center-pivot irrigation systems to the tune of around $130 per acre through the system of investment credits and capital gains taxes now in effect. If we view the recent history of cropland development without taking these public subsidies into account, we are seriously miscalculating the ease with which new land can be put into crops.

The summation of all these factors does not prove the economists totally wrong, but for the most part I find their analyses far too complacent. Future expansion of the land supply will run into limits, and could operate far too slowly to prevent significant dislocations in U.S. agriculture. Public policy needs to create a "safety net" that precludes the permanent loss of either productive farmland or skilled farmers.

Needed: Long-range political solutions

The next problem facing the Nation as it tries to create that safety net may not be the financial or environmental crunch affecting the land but the political stampede to do something about the situation. That the issue will come to the forefront is not so much in doubt; that responses will be devised seems equally certain. That they will be misdirected is highly likely.

The reaction of public policymakers to long-range natural resource problems is, unfortunately, likely to be short-run in nature. Again, to quote Dennis Meadows in his assessment of the energy situation:

"Two policy options are open to the federal establishment. The first causes energy prices to decline over the short term but leads to rapid escalation of prices later on. The second increases prices now but lays a foundation for price stability later. Most forecasters overlook the fact that the federal establishment will systematically pick the first policy over the second so long as the most difficult consequences only begin to manifest themselves after the next election."

We see clear indications of that tendency in the agricultural arena. Despite the financial disaster imposed on American farmers by pushing foreign grain sales at prices well below production costs, the official USDA policy is to increase sales abroad. Efforts to establish a means of holding out for a fair price are rejected for various reasons, most of which have to do with a fear of injecting more federal control in the marketplace or encouraging foreign countries to expand agricultural output. While all these arguments have merit, they overlook one thing: if the current method of doing business bankrupts many farmers or forces them to ruin the land, it may be the worst choice of all.

Selling more than can be produced within the limits of sound land and water management creates a treadmill where the more we grow, the lower we drive prices, and the more land we damage. As the land's productivity is damaged, it costs more to produce, so we are forced to try to produce still more, at still lower prices, to try and pay the rising costs of production. It is a losing game, where we are forced to run faster and faster just to stay even.

The answer to the land damage created by this dilemma does not lie in new or different federal conservation programs, but in developing a coordinated market/conservation system that provides price levels adequate to repay not only the costs of production but also the necessary reinvestments in the land that will conserve, protect and enhance its long-term productivity. Methods and institutions for doing this exist, but so far organized agricultural groups and national farm policymakers have been unwilling to risk the political wrath of agribusiness (and many farmers) by imposing such a system.

Creation of a marketing system that rewards farmers who choose to apply appropriate soil and water conservation management must fall to the federal government. It is only at the federal level that marketing programs have any chance of functioning at all. But in the age of "new federalism," where the conventional political wisdom is to get the central government out of activities rather than into new ones, this idea is definitely out of phase. So, in spite of the fact that most international food buyers and sellers are governments rather than individuals, the United States is reluctant to move into this arena.

Identifying the farmers who are carrying out appropriate management systems on their land is a job that is too localized for the federal government, but the combination of state and local action agencies created by the soil-conservation district system could handle the task. In Iowa, for example, state law sets up a system of allowable soil-loss standards, which are then refined and administered at the local level through conservation districts. Few other states have followed this lead, however, because it introduces difficult political issues when local citizens begin to administer programs with real impact on their neighbors. Again, we too often opt for the long-range disaster to avoid the political consequences of taking needed actions.

In the long run, of course, we must establish a policy framework in which a sustainable agriculture can emerge. A sustainable agriculture is one that produces the food, fiber, energy and other crops that the nation needs, including a marketable surplus that can be sold abroad. It produces this on an average year, not just during times of good weather. It weathers a bad year by drawing on stored fertility and moisture in the soil; stored water in reservoirs; stored wealth in financially secure farms; and stored food products in the granaries of farmers, industries and government. It profits from a good year by setting aside extra commodities or making an extra effort to see that they are sold abroad, without driving prices though the floor and creating financial hardship or ruin among producers. In addition to meeting domestic food needs and a substantial export market, a sustainable, regenerative agriculture could reduce the waste and pollution of water, provide better wildlife habitats, slow the advance of desertification and soil salinization, reduce the loss of prime farmlands and fragile topsoils, and, in general, make rural America a far more healthy, satisfying and financially rewarding place to live. It is a utopian goal, perhaps, but an essential goal nonetheless.

This will require the nation to establish production levels — along with accompanying agricultural technologies — that can be indefinitely sustained with the land, water, technology and capital at hand. It will mean a concerted effort not just to develop new lands or technologies but also to carry out the needed maintenance and upgrading that existing resources require. In economic analysis it may be inconsequential whether we spend dollars on new development or spend them on maintenance of existing systems. From the political standpoint, it is often far more attractive to promote new development. But in terms of real economic and social progress, we will be limited by our willingness to conserve and maintain today's farmlands and the farmers on it.

Converting today's depleting agriculture to a new

type of regenerative agriculture will not be a quick or painless process. But some things can be identified as starting points:

• We can take immediate action to overhaul national commodity programs. We could double efforts and incentives for soil and water conservation, which today attract less than one-tenth of one percent out of each dollar in the federal budget.

• We can encourage the development of soil and water conservation practices such as no-till or minimum-till farming that result in both soil and energy savings, and we can do it without resorting to indiscriminate pesticide usage, which can negate all the cost savings involved and replace soil erosion and sediment pollution with chemical pollution — a trade nobody wants.

• We can create ways to help ease farmers off of lands that are too marginal for sustained agriculture, and eliminate the current financial incentives to attack such lands with the plow.

• We can set up an intermediate financing bank to prevent a massive wave of bankruptcies created by the debt spiral in which farmers are currently caught. The bank could offer long-term loans at subsidized interest rates so that farmers could refinance their debt in a manageable plan. Those farmers who accept low-cost loans from the federal government should also accept a requirement that they manage their land within local conservation standards. If they are unwilling to make that commitment, the public should be unwilling to keep them in business at public expense.

• We can accelerate research and extension work in development of on-farm techniques that help farmers wean themselves from the industrial model that has them on the current cost-inflation treadmill. That would include techniques like crop rotations, organic farming techniques, integrated pest management and conservation tillage. These techniques often have too little commercial potential to be attractive to industrial research and advertising campaigns, so they should be the prime target of public research in USDA and the agricultural universities and experiment stations.

• Finally, we can be more aggressive at demanding, as a people, a more sensitive use of the land and water that sustain life. When an industrial user bulldozes up an acre of prime farmland for a strip mine, or covers it with concrete for houses, or poisons it with toxic wastes, he has taken an action that is more than private. He has robbed countless people — for countless generations — of an essential part of their opportunity for a constructive life. Such behavior is immoral, and it is time the American people made it plain that it can no longer be tolerated. We know how to do it better, and now it is time to show the political courage to do what must be done.

2. State-Sponsored Research Program on Population Exposure to Toxic Substances

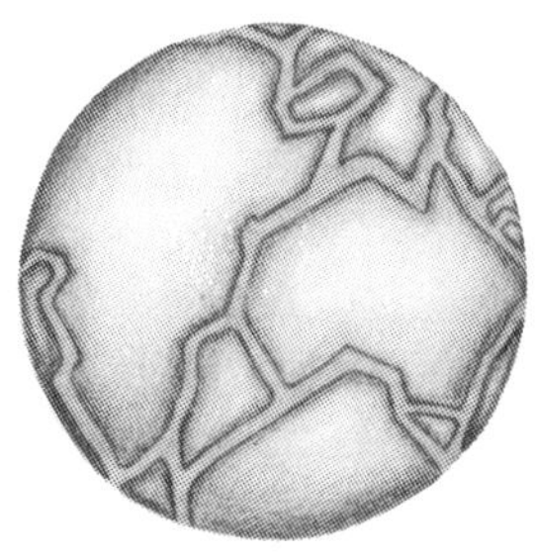

"We cannot, in the short run, power automobiles, paint houses, grow food, make and dye clothes, or print books without generating substantial amounts of toxic wastes."
Epstein, Brown and Pope.
Hazardous Wastes in America;
Sierra Club Books, 1982.

Dr. Michael Greenberg
Director
Public Policy and Education Division
University/Industry Cooperative Center for Research in Toxic and Hazardous Materials
Rutgers University
and
T. Burke
J. Caruana
G.W. Page
K. Ohlson
Published in 1981
Volume 1, Issue 1
The Environmentalist

2 Population exposure to toxic chemicals in the environment has become one of the most important, if not the most important environmental issue of the 1980s in the United States. In 1976 residents of New Jersey were alarmed to learn that between 1950 and 1969 New Jersey had the highest white male cancer rate in the country. The white female, non-white male and non-white female rates were also among the highest. In response, Governor Brendan T. Byrne (under Executive Order #40, May 26, 1976) directed the departments of Agriculture, Environmental Protection, Health, Higher Education, and Labor and Industry to develop plans for the prevention and control of cancer. An unprecedented budget for a state government, \$1.2 million dollars for 1978, was allocated. Other funds were added from federal sources. Descriptions follow of parts of the extensive research program that has been established.

The Department of Environmental Protection created a Program on Environmental Cancer and Toxic Substances to try to learn more about, and to develop and implement solutions about, environmental factors that appear to be associated with the state's high cancer rates. The research component of the program has had two major thrusts. One is aimed at finding indirect evidence of the association of cancer mortality and risk factors. It involves finding rates of trends in cancer mortality and relating these to risk factors. More specifically, cancer mortality trends from 1950 to 1975 have been studied for the New Jersey-New York-Philadelphia metropolitan regions and the remainder of the United States for 65 age-, sex- and race-standardized causes of cancer mortality. The goal has been to determine if the region exhibits unique patterns or a profile similar to the remainder of the nation. In addition, in order to generate clues for epidemiological research, the geography of cancer mortality for 24 age-, sex- and race-standardized causes has been compared to the geography of 71 elevated and reduced risk factors over the 49 counties in the study area within New Jersey.

The second thrust aims at finding direct evidence of exposure by studying the presence of toxics in the water, air and at production sites. The second is the focus of this paper. Particular attention is focused on a set of studies that are part of the larger state effort. Specifically, the paper presents three approaches and initial results of three studies:

(1) population exposure to toxics through water;
(2) population exposure to toxics through the air;
(3) the industrial survey and the association between toxic pollution and industrial activity .

The three direct evidence studies are addressed for two reasons. First, the first group of cancer mortality studies has been widely publicized; the second group has not. Second, although seemingly less newsworthy than cancer patterns, it is from careful monitoring and reflection on the level of environmental contamination and potential population exposure that government will be able to act with firm conviction to protect public health and the ambient environment.

The three studies are presented in order of their accomplishments to date. The water studies are the most advanced, having been under way for more than three years. In contrast, the industrial survey has begun to yield results for less than one year. To the best of our knowledge the state of New Jersey is in the forefront with respect to each of these three projects.

Population exposure to toxics through water

The last decade has witnessed a great deal of speculation concerning the relationship between water quality and cancer. Numerous papers have reported the presence of confirmed or potential carcinogens in surface waters. One literature review reported that 1,259 different chemicals have been found in water sampling. Probably the most noteworthy report was issued by the Environmental Protection Agency in 1975 when they reported finding more than 250 different organic chemicals in U.S. drinking waters.

These findings may be frightening, but they are all based on short-term sampling programs. These samples may or may not indicate widespread long-term contamination. In addition, more than two dozen studies have found statistical association between cancer mortality rates, usually urinary and digestive tract, and possible indications of toxic contamination of water supplies.

Does the water pose a threat to the public?

To answer this question for their waters, the state of New Jersey has embarked upon an unparalleled monitoring program. The initial questions were what parameters to sample, where and how often should they be sampled, and what types of tests should be made?

There is an established literature and Environmental Protection Agency guidelines for anyone interested in establishing a water-quality baseline for standard water quality parameters. The central tendencies and deviations of dissolved oxygen, suspended solids and most of the standard parameters are reasonably well-known owing to thousands of sampling studies. Accordingly, the number of samples to achieve a desired level of confidence in the study can be predetermined.

At the time the studies began there were no guidelines or even major sets of samples to consult for carcinogens. Indeed, to the best of our knowledge, the data used in this paper represented the first reported use of new analytic methods to detect extremely low concentrations of many toxics and carcinogens.

Given the restrictions of a limited literature, a developing sampling technology, and a cost of between $300 and $1,000 for each sample, an extremely conservative sampling approach was adopted. It was decided to focus initial attention on groundwater because it is thought to move slowly through the environment. Thus, it was felt that higher concentrations would be found and more readily associated with surrounding land uses than was likely to be the case with surface-water samples. The initial data were 408 groundwater samples analyzed for 45 substances — light chlorinated hydrocarbons, heavy chlorinated hydrocarbons and heavy metals. The existing wells chosen as test sites represent a cross section of locations, and land and water uses. About 20 were chosen from each of the 21 counties involved in the study. Nearby land uses include industrial, commercial, urban, suburban, agricultural and open space. The wells are used for agriculture, cooling water, industrial processes, monitoring of land uses like landfills and testing of potable water supplies. In addition to the 408 samples, about 10 percent re-samples were taken in order that we would have some idea of the accuracy of the equipment.

Nearly all of the chemicals were found in at least one well. More than 95 percent of the wells had extremely low concentrations of the substances. The typical (model) value for almost every organic substance was zero, the typical value for the metals was usually the minimum concentration that could be detected by the chemists. While most of the contamination is quite low, a few values are high. Seventeen of the highest values exceed 10 parts per billion (ppb) and five exceed 100 ppb.

Are the concentrations high enough to warrant concern?

This question cannot be answered with certainty. However, an idea of the order of magnitude of the threat for some of the chemicals may be obtained by combining the potable groundwater samples (312 of 408) and EPA estimates of cancers produced by consuming ambient water over a lifetime. Specifically, the EPA has published water-quality criteria for 65 specified toxic pollutants. The EPA health effects model is a very conservative linear, no-threshold and dose-response model. The model assumes no safe dose — one molecule can lead to cancer. As the dose increases, the probability of getting cancer increases. The model is applied to an average-sized American male who drinks an average amount of water and eats some fish.

We applied the EPA model to 26 of the chemicals for which we had data for the 312 samples. More than half of the groundwater samples, if consumed for a lifetime, are estimated to cause less than one additional cancer per 100,000 water consumers. Eighty percent of the samples would be estimated to cause four or fewer cancers. The t3 percent most contaminated samples are estimated to cause more than 50 additional cancers per 100,000 consumers. The worst samples (highest percentile) could cause more than 500 additional cancers. This exercise, while providing a preliminary order-of-magnitude estimate for some water contaminants (since one out of three Americans is likely to contract cancer during their lifetime), cannot be taken at face value because the results are dependent upon numerous assumptions many of which are highly debatable.

Further questions:

Beyond trying to seek initial answers to the questions of what substances are present, in what concentrations are they precent, and how potentially harmful are these concentrations, the answers to four other initial questions were sought:

• Which pollutions are found in conjunction?

• Is there an apparent relationship between land use and the geographical pattern of contamination?

• Can a few pollutants be used as surrogates for a larger number of pollutants?

• Can one distinguish between wells contaminated at background levels and grossly polluted wells?

Statistical grouping methods were used to determine which chemicals were found in conjunction with others.

Few significant associations were found between chemical groups. Most of the strong associations were within chemical groups. Specifically, two pesticides, one light chlorinated hydrocarbon and one heavy metal group were found. By and large we observed the expected chemical associations. For example, DDT, DDD, DDE, aldrin and endrin which are metabolites were often found together. (Aldrin is oxidized to dieldrin when exposed to light or in the bodies of organisms.) In addition, similar chemicals have been used to control the same insects. It is not surprising to find them in the same place. Finally, the manufacturing process introduces impurities into the chemicals. For example, heptachlor is a common impurity in chlordane. Overall, we found strong associations within groups of chemicals and weak association between these groups.

Local land-use does seem to be related to some types of groundwater contamination

The 10 samples most heavily contaminated with pesticides were identified, as were 10 for light chlorinated hydrocarbon and heavy metals. A fourth group was 10 samples virtually free of any of the toxic substances. Land-use information was collected for the 10 miles surrounding each of the 40 sample sites. Statistical analyses revealed that gross pesticide contamination was found in areas with agricultural activities and forest within one mile of the sample sites. Most of these sites are in rural, southern New Jersey. Gross light-chlorinated hydrocarbon pollution was found in areas with many commercial and industrial activities within one mile. Most of these sites are in the northern urban corridor. The land-use profiles of the heavy metal and clean sample sites were not distinct.

Could small parameters be used to infer a larger set?

At a cost of $300 to $1,000 per sample, substantial savings could be realized if standard water-quality parameters of a few carcinogens or toxics could be used as a screening device to select sites for initial or follow-up monitoring. A statistical method was used in these efforts. No success was found in correlating the standard water-quality parameters and the 45 toxics. Some success was realized in finding associations between a small and large set of pesticides. Four pesticides (endrin; p,p'-DDD; lindane; and heptachlor) were able to predict the values of 10 other pesticides. However, the fits are not strong enough and the data not substantial enough to warrant anything but optimism to continue the research into finding surrogates as more data become available.

Do the chemical concentrations in the groundwater represent gross pollution or normal concentrations?

This project is one of the first steps by the state of New Jersey in establishing standards to protect groundwater from pollution, abate existing pollution, and meet the requirements of the interim primary drinking water standards established under the Safe Drinking Water Act of 1974. The initial method is a probability-plotting procedure using either normal or logarithmic-normal probability paper. The concentrations of the substances are measured along the vertical axis, the probability along the horizontal axis. If two populations exist, then two distinct lines should be found. The lower linear segment will be the normal background sample population, the upper linear segment the grossly contaminated segment.

The method generally produced the expected two populations. For example, for cadmium, 383 samples fell within the 0.5 to 10.0 ppb range. The upper curve contained three points ranging from 42 to 402 ppb. The land uses surrounding the three wells were consistent with a local land-use pollution cause. Sometimes the transition from the normal background to the grossly polluted was blurred. The graphical judgment method has been replaced by statistical outlier methods. Case-by-case and block outlier methods are currently being evaluated. The ultimate goal is to develop the basis for non-degradation standards.

Continuing research

Based on the results of the initial studies, an expanded set of locations and chemicals are being tested. The state of New Jersey has now collected over 1,200 well and 700 surface-water samples. The chemists will expand the tests to include new substances as well as those tested for in the initial studies. In addition, to research being conducted by the authors, other studies are under way. Biological studies have been added. The biological studies consist of applying the sample to bacteria for evidence of induced mutation. Based on nearby land uses, the samples should represent a range of water conditions. They will be taken from groundwater, surface water, sediments in water bodies and effluent, raw and tap-water sources. Monthly and seasonal tests will be made on a set of wells in order to ascertain trends. Finally, as part of a joint New Jersey and National Cancer Institute project, all potable water purveyors serving over 1,000 people are being tested for volatile organics and heavy metals. This will provide a measure of drinking water quality for over 95 percent of the state's population.

Two general conclusions drawn from this research

First, monitoring programs can pay immediate dividends. The state of New Jersey monitoring program has identified contaminated water sources and sources of contaminating substances. The state government has initiated actions in response to the monitoring, ranging from the closing of some sources to legal actions to clean-up programs. Second, detailed statistical analyses of the data reviewed above imply a need to refrain from oversimplified characterizations of the extent of groundwater contamination, of sources of pollution, and of the association between water pollution and rates of disease.

P*opulation exposure to toxics through the air*

Like the relationship between water quality and cancer, the relationship between air quality and cancer is hotly debated. In 1932 Tobey claimed that: "Endeavors have . . . been made to show that smoke may have some influence on cancer, but there is actually no reliable data to incriminate smoky atmospheres, undesirable as they may be from the standpoint of general hygiene."

Four decades later, a National Academy of Sciences

Conference on the Health Effects of Air Pollution (1973) concluded: "There are substances present in polluted atmospheres which have been demonstrated in experiments on animals to be mutagenic or carcinogenic. The concentrations required in such experiments have been very much higher than those which might be encountered in polluted atmospheres. There is an urban-rural gradient in human lung cancer not attributable to cigarette smoking. The atmospheric pollutants here considered may be among the factors contributing to the higher rate in urban areas, but definitive evidence is lacking."

The literature follows a tortuous path to arrive at the latter conclusion. Studies that link air pollution to cancer abound. These studies make a case for air pollution as a cause of cancer by producing evidence that an urban/rural difference in lung and sometimes other cancer exists and that smoking could not account for the difference. Air pollution was implicated as the missing agent with little direct evidence.

Like water contamination, direct evidence for carcinogens in the atmosphere began to be gathered in the early 1960s. Most of the early attention has been devoted to organics.

Initial questions:

It is difficult to know the destination of pollutants released into the atmosphere because, unlike the land and water media, air can move in any direction. Accordingly, the first two years of the New Jersey research have concentrated on establishing a baseline from existing data sets and gathering an initial set of data on toxics and carcinogens.

With respect to standard pollutants, particulates were chosen for a baseline over carbon monoxide, oxides of nitrogen and sulfur oxides for three reasons. More than three times as many readings are available for particulates than for any of the other pollutants. Second, components of some particulates are directly carcinogenic, others are thought to be carcinogenic. Finally, it is possible to recover the particulate samples for biological testing. The last capability is especially noteworthy.

Choosing carefully among the many thousands of particulate samples, the analyst can learn if specific locations and levels of particulate air quality are associated with increased risk of mutation and, therefore, possible exposure to carcinogens.

Particulate air quality data were available for New Jersey for 1966-1977. Prior to 1973 the data are difficult to use for a statewide baseline. The station locations frequently were changed. A large proportion of the stations were located in urban/commercial areas, and the height of the samplers varied widely.

By 1973 these weaknesses had been corrected. Seventy-four samplers located throughout the state are suitable to construct a baseline. These stations collect the data with high-volume samplers. The samplers, which function like vacuum cleaners, suck in particles for 24 hours every sixth day. The machines are accurate to within plus or minus 5 ug/m3. The stations are located between 10 and 20 feet above the ground. About half are in New Jersey's urban-industrial corridor, the other half in exurban and rural areas. Overall, the 74 samplers all had at least nine full months of data for each of the five years of interest.

The data base they offer is substantial. The record contains almost 18,000 particulate samples. The results are quite striking. Particulate levels in New Jersey display patterns which seem to be related to land use and atmospheric transport. Land use is the most influential factor. The lowest annual geometric averages (25-30 ug/m3) are found in sparsely developed northwestern and southern New Jersey. Agricultural areas have relatively low levels of particulates with annual geometric averages ranging from 30-40 ug/m3. Suburbs with densities of at least 1,000 persons per square mile can be identified by their annual geometric averages of 40-50 ug/m3. Most of these suburban communities fall within a funnel-shaped corridor running from Philadelphia and Camden in the southwest, to New York City in the northeast. Samplers which consistently exceed the primary and secondary standards for annual geometric means are found in the concentrations of industrial, commercial and highway uses which characterize the innermost ring of intensive development in the New Jersey-New York City metropolitan areas.

Land activities associated with the work week, particularly the journey-to-work trips, affect particulate levels in a predictable weekly cycle. Thursdays and Fridays are typically 10 ug/m3 higher than average, whereas Saturdays and Sundays are 10 ug/m3 lower.

Widespread pollution episodes seem to be associated with unusual meteorology and emission factors such as fires, droughts and wind patterns which blow particulates back and forth across the state without permitting dispersal. Episodes of low particulate levels are usually associated with at least several days of precipitation and winds from regions with low particulate levels.

Does it follow that high particulate episodes and high daily averages imply increased exposure to carcinogenic and cocarcinogenic substances?

The second phase of the particulate study is seeking answers to these questions. We have selected the best and worst days and stations on these days. It is hoped that the biological tests at these stations will shed some light on these extremely important questions.

Biological testing will not be the only source of information. Chemical monitoring for 17 toxics was begun during the second year. The initial 100 metal and 300 organic air samples, like the water sampling, required the use of pioneering chemical methods, because no governmental agency has set forth a standard procedure for measurement of these substances at the part per trillion level. Indeed, few laboratories in the United States are capable of performing the analyses for chemicals in our studies which include Lead, Zinc, Arsenic, Manganese, Nickel and Mercury, and organics like Vinyl chloride, Chloroform, Benzene, Carbon tetrachloride and several others.

Unlike the water samples, the air sites are not representative of all parts of the state or of the spectrum of land uses in the state. Rather, they are concentrated in three areas: (1) an industrial area in Newark; (2) a residential area in northern New Jersey (Rutherford) that witnessed an episode of high childhood leukemia and Hodgkin's disease; and (3) industrial, commercial and residential areas in central New Jersey.

Like the water data, low levels of contamination were detected in almost every sample. The typical values for seven of the 11 organics was zero, while for three of the other four the minimum detectable value was the mode. The lone exception was benzene. It not only was ubiquitous, but the modal value was 1.4 ppb. This was not surprising because US/EPA data consistently indicate relatively high levels of atmospheric benzene, especially in urban areas. Again, as in the water data, the highest values usually exceed 1 ppb. Specifically with respect to the organics, the 90th percentile values of seven of the 11 chemicals exceed 1.5 ppb, and 10 of the 11 highest values exceed 10 ppb.

The above pattern is replicated among the metals, though the metals are found more often than the organics. Lead and nickel, like benzene among the organics, were found in virtually every sample (99 percent). The others were detected less often: arsenic 53 percent; manganese 81 percent; mercury 88 percent; and zinc 58 percent. The greater detection of metals than organics is probably due to their presence in bedrock.

Are these air concentrations enough to warrant concern?

There are no EPA dose-response models to use, although there is an ambient air-lead standard. If judged relative to threshold limit values recommended by the Occupational Safety and Health Agency for workplace atmospheres, the ambient atmospheric numbers are several orders of magnitude less. However, the workplace and outdoor environments are not equivalent. Accordingly, one cannot assess the implications of the existing data at this time.

Further questions

Given the limitations of the initial toxics data, additional analyses were limited to three questions:

- Which pollutants are found in conjunction?
- Is there a regional/local pattern of contamination?

• Does pollution vary by time of day, wind speed, wind direction, temperature and weather conditions?

Statistical grouping methods helped answer the first two questions. The 11 organic chemicals formed three groups. One group, named the ubiquitous urban chemical group, consisted of sampling sites which manifested high readings of 1,2-dichloroethane; 1,1,1-trichloroethane, benzene, trichloroethylene and tetrachloroethylene. These five substances are all extremely widespread. Sources include automobile exhaust, gasoline stations, chemical manufacturing, petroleum refining and solvents. Most of the high readings were found in the two northern New Jersey areas.

The second group revolved around three chemicals containing benzene: P- and O-dichlorobenzene and nitrobenzene. The location was quite distinct — the northern New Jersey industrial area. The use of these three chemicals is restricted in comparison to the five that formed the first group. These substances are probably manufactured together in the northern New Jersey industrial area.

The third group focuses on carbon tetrachloride and chloroform in central New Jersey. The common link is again probably local manufacturing, because both substances are manufactured by direct chlorination of methane.

Little consistency was observed among the six metals. The major locational finding was a strong tendency for high lead readings to be near major highways. This is not a surprising finding given the fact that 90 percent of atmospheric lead is discharged by automobiles. Lead smelting and refining, storage battery production, and paint manufacture are the other important sources. The other metals did not show an obvious pattern, but did tend to be higher in the denser urban areas.

While chemical consistency and location patterns were manifested, there was little consistency by time of day, wind speed, wind direction, temperature and weather conditions. Unstable weather conditions seemed to be associated, though not consistently, with elevated levels of some organics.

Continuing research

The air monitoring research has not been under way for a sufficiently long period to lead to important government actions or to research conclusions (as of 1981 when this article was first published). Nevertheless, it has yielded testable hypotheses about the relationship between ambient air quality and sources of toxic substances. Present work focuses on a wider set of substances for a limited number of sites. The samples are being taken at six-day intervals to coincide with particulate monitoring sampling. While the sites are primarily urban, one rural site has been included as a control. Future work will expand the number of monitoring sites and focus on the temporal sequence as well as the spatial distribution of toxic substances in the outdoor air.

Survey on toxic pollution and industrial activity

The water and air analyses have begun to identify places of high- and low-population exposure to toxics. Some have led to the identification of sources of contamination. In time, these findings may lead to the identification of many sources of toxics. Without doubt, however, a survey of large point sources is necessary. Currently, no such survey exists. Under provisions of the Federal Water Pollution Control Act of 1972 and its 1977 amendments, data on about 80 pollutants were obtained from industry. Relatively few of the 80 are known or suspected carcinogens. Similar spotty data are also available about air emission sources. The Toxics Substances Control Act of 1976 requires industry to submit data on toxic substances to the EPA. However, implementation of TOSCA is proceeding slowly, and the range of information likely to be requested by EPA and industrial coverage will probably fall short of the needs of states in the Middle Atlantic Region, and, in particular, New Jersey. Measured in dollar-value outputs, New Jersey produces more chemicals than any other state in the United States. It also has major metal fabricating and petroleum refining sectors. Data concerning the use of carcinogenic and toxic agents and disposal methods are inadequate.

Initial activities

Therefore, in 1976, the N.J. Departments of Environmental Protection and Labor and Industry jointly began to develop a survey form to provide information on industrial sources of carcinogens and toxics. The survey sought information on the use, storage, release and disposal of 155 substances. Industries are to furnish the amounts of the substances they use and their air, water and solid-waste emissions.

The survey has taken two routes. One is administrative. The original questionnaire was circulated among agency and industrial representatives. The second route is technical. The original and modified survey forms were designed keeping the computer in mind. After a pretest which is currently underway is completed, the intent is to inventory between 10,000 and 15,000 factories.

The enormous potential size of the data bank and the confidentiality of some of the data led us to carefully evaluate different data-base management systems. We chose RAMIS, which allows the user to manipulate, store, retrieve and display large quantities of data on an IBM system. One of its important advantages is that it can be invoked through either English-like languages or the traditional procedural languages (e.g., FORTRAN, COBOL and PL/1). A second advantage is that the data are stored only once. Data are stored in a hierarchical structure, and information can be drawn from several files for one report. A third strength of RAMIS is that it easily generates reports and various graphical outputs. Fourth, RAMIS allows the user to preserve the confidentiality of data.

The flexibility of RAMIS allows us to generate at different levels of detail. For example, the state may wish to know the output of a specific chemical for every factory in the state. They may also wish a complete report on all 155 chemicals of an individual company. Again, the state may wish to have county or river basin summaries. Finally, RAMIS has a relatively simple mechanism for updating the files and adding new files.

Continuing research

There is a general plan for coordinating the industrial survey results with the water and air data reviewed in the previous sections. The goal of the plan is to determine if there are consistent associations between the spatial distributions of the substances emitted into the environment and contamination of the environment. That broad goal requires the merging of the air, water and industrial survey data sets and locating each source and link. Then, while controlling for other land uses, statistical methods will be used to measure any associations. Pilot studies are currently under way.

What has the New Jersey experience taught us?

The question which initiated this research was why did New Jersey have the nation's highest cancer mortality rates during the 1950s and 1960s? Since it has been found that the remainder of the United States has been rapidly catching up to New Jersey in cancer mortality, the emphasis has shifted to monitoring the ambient environment and reducing population exposures.

The research program, only part of which is summarized here, brings to bear some pioneering and some standard applications of chemistry, data management and statistical analyses. We have opted for conservative research designs in an effort to determine if statistical tools can describe relationships expected from chemical, biological and land-use theories.

The state-sponsored research program has brought together scientists with a wide variety of training and experience. The disciplines represented by the program include geography, meteorology, epidemiology, toxicology, statistics, demography, biology, chemistry and computer science. Unlike many "bench" science investigations, environmental evaluation requires an interdisciplinary approach. This has created a need for new and modified approaches to the training and continuing education of environmental scientists, i.e., training programs that introduce the scientists to the wide range of disciplines involved in environmental decision-making, while at the same time allowing cultivation of an area of specialization. The complexities of evaluating the impact of environmental pollution have also increased the needs for personnel training in auxiliary fields ranging from economics to public relations.

The state program has also highlighted the need for effective communication among scientists, public officials and the general public. In this case the program was made possible by a heightened citizen awareness due to the release of New Jersey's cancer mortality statistics. The concern of the public is essential to the successful continuation of state-sponsored research. Political leaders and the public expect answers. Scientists must be able to explain the results of environmental research in layman 's terms. They must set practical, achievable goals and dispel false expectations. In addition, researchers should be available to discuss investigations with the press to assure that goals, methods, and results are accurately reported. As early as 1980 it was suggested that intensive courses should be offered to better prepare the media for reporting complex environmental research. [Editor's Note: By 1985 pilot educational programs for members of the media were being conducted at the Institute directed by Dr. Greenberg. These programs are described briefly in Chapter Three.]

Communication and cooperation among state governmental agencies is also needed for successful projects with the EPA and the National Cancer Institute. Recently an agreement was signed by the New Jersey Department of Environmental Protection Department of Health and Department of Labor and Industry to coordinate environmental cancer control efforts. The maximum benefits of environmental research can be realized only when interagency competition for limited resources and duplications of efforts are minimized, and diverse talents and resources are joined in cooperative efforts.

3. Transboundary Problems Between the United States, Mexico and Canada

Marshall E. Wilcher
Assistant Professor,
The Pennsylvania State University
Published in 1983
Volume 3, Issue 1 of
The Environmentalist

3 Among the many disturbing, perplexing and crucial aspects of the environmental crisis are the problems resulting from transboundary pollution. These problems have the potential, of course, for affecting the United States on both its northern and southern borders. Looking first to the south, relations between the United States and Mexico take place in an historical context marked by war, intervention, and cultural and economic penetration. Because of this, relations in any issue area are subject to being hampered by mistrust and suspicion. Cooperation during World War II enhanced relations between the two countries. Although the U.S. administrations since the Kennedy years have attempted to create the nature of a special relationship with Mexico, the record has been spotty and uneven. Several important issues complicate U.S.-Mexican relations in the 1980s. The most important of these are international trade issues, energy issues, the migration problem and border relations. Mexico is the third largest trading customer of the United States, and the United States is the greatest source of Mexican imports.

The vast oil and gas reserves of Mexico and the proximity of the United States as the most logical importer of the resources dictate that energy and trade issues are to be an important and permanent element in U.S.-Mexican relations. These relations were damaged by the prolonged negotiations on natural gas, which were finally completed with an agreement which was expected to reach 300 million cubic feet per day in the 1980s. Further, the United States accounted for 80 percent of Mexico's oil export in 1979. Mexico hopes to diversify its oil-exporting trading partners, but the oil and gas exports to the United States are logically and likely expected to continue to grow in volume. Although there has been a dramatic increase in Mexican exports to the United States, U.S. exports to Mexico increased even more. In 1981 Mexico's trade deficit with the United States approached $3 billion. This situation is likely to continue and to complicate general relations.

Likewise, the migration issue is likely to continue as an irritant to relations between the U.S. and Mexico.

The trade, energy and migration problems are potentially more disruptive to relations than transboundary environmental problems. Because of this, efforts to deal with transboundary disputes can be complicated by lack of progress in the more troublesome issues.

U.S.-Mexican cooperation on environmental issues

The nations do have a history of cooperation in transboundary environmental problems. The oldest organization for dealing with boundary problems is the International Boundary and Waters Commission. The forerunner of this agency dates from the late 19th century, based upon the provisions of a treaty on border waters between the two countries. The IBWC is responsible for:

- The distribution between the two countries of the waters of the Rio Grande and the Colorado Rivers;
- The regulation and conservation of the Rio Grande by international storage dams and reservoirs for the utilization by the two countries within the allotments specified to each;
- The protection of lands along the rivers from floods, by levees and floodways;
- Preservation of the Rio Grande and the Colorado River as the international boundary;
- Demarcation of the land boundary;
- The preferential attention to the solution of border sanitation and other water quality problems.

The commission also operates and maintains some international water control structures along the border, detects potential problems within the jurisdiction of the commission, and consults with local, state and federal domestic authorities in carrying out its work. The commission is made up of U.S. and Mexican sections, located in El Paso and Cuidad Juarez, and is primarily a technical- and engineering-oriented organization. Much of its work entails fostering international cooperation through national agencies of the two countries. The IBWC has been involved in numerous border-related problems dealing with the levels, flows, irrigation, distribution and conservation of the waters along the border. Formal agreements relating to environmental issues date from 1936, when a Convention for the Protection of Migratory Birds and Game Mammals was signed. In 1972 the nations agreed to establish a Joint Air Pollution Sampling Project to gain information on air pollution in the El Paso-Cuidad Juarez region. During that same year the nations agreed that the United States would undertake measures to reduce the salinity of the waters of the Colorado River delivered to Mexico.

At a meeting between President Carter and President Lopez Portillo in 1977, an umbrella organizational arrangement for mutual relations was established to give added significance to the relationship between the two countries. This arrangement was known as the U.S.-Mexico Consulting Mechanism, and was organized into eight working groups focusing on energy, trade, finance, industry, tourism, migration, border cooperation and law enforcement. The groups were to meet at least annually and more often if necessary. The groups were headed by representatives of the foreign ministries, but included representation from various national functional agencies from both sides of the border.

Environmental problems along the border are primarily related to air and water pollution along the Rio Grande and Colorado Rivers and in the 16 Mexican and 13 American communities along the border. With the formation of the consulting mechanism there was an increase in transboundary environmental cooperation. One striking feature of this cooperation was the scale of joint activity and interagency cooperation carried out between the United States and Mexico. For example, the U.S. Environmental Protection Agency engaged in training programs for Mexican government representatives, carried out extensive information exchange, and facilitated some technology transfer. Although care must be taken to avoid a patronizing posture on the part of the United States, the government of Mexico has been most interested and has participated in several of these exchange programs. Clearly, appreciation of the general goals of environmental protection and concern over the environmental crisis is not as great in Mexico as in the United States (as is the case in most lesser developed countries). As one Mexican scholar expressed it, "for Mexico, the issue is not a question of convenience . . . it is a question of development and life itself." He further points out that Mexico's pollution results from poverty, and can be solved only by development. It should be noted that the U.S.-Mexico border is quite distant from the central economic zone of Mexico and Mexico City, and that region may consume many of the resources devoted to environmental protection. Yet Mexico is obligating some resources to specific pollution problems in the urbanized portion of the border.

Along with this general atmosphere of intergovernmental and interagency cooperation, some progress appears possible in concrete achievements aimed at improving air quality and water quality along the border. Air pollution along the border results from scattered industry on the U.S. side, cooking fires and open burning of garbage on the Mexican side, agricultural operations on both sides, and automobile emissions at border crossing points. In 1978 the U.S. Environmental Protection Agency and its Mexican counterpart agreed to a Memorandum of Understanding on air pollution and the transport of hazardous wastes.

The thrust of joint efforts has been in sampling, measurement and monitoring. Working in conjunction with local and state authorities in El Paso, Texas, and Cuidad Juarez, Mexico, and utilizing largely EPA resources, the governments have established a joint air-quality monitoring program and hope to compile data for further action.

In May 1983, EPA Administrator Ann Gorsuch visited Mexico and agreed to undertake similar arrangements in the San Diego/Tijuana metropolitan area. The EPA and the Mexican government have also been involved in a joint monitoring project designed to compile a health hazards inventory in the urban areas along the border.

Water quality problems generally involve sanitation and result primarily from sewage discharged into the Rio Grande, Colorado and New Rivers, and in ocean areas around the San Diego-Tijuana metropolitan complex. Both governments are committed to improving the water quality in ocean waters around San Diego and have taken some concrete steps toward that end. In addition, the governments agreed in 1980 that the International Boundary Waters Commission should be assigned the task of planning and monitoring measures to alleviate several areas where Mexican wastes are polluting waters entering the United States. Mexico has taken interim steps to reduce these pollutants in each of these areas until permanent arrangements are agreed upon. In one of these areas — the New River at Calexico, California, and Mexicali, Baja California, — an agreement was reached calling for a specific permanent solution.

A troublesome transboundary environmental problem, which has a history of exacerbating U.S.-Mexican relations, is the salinity content of the Colorado River. The problem dates from 1960 when it was noted that the salinity level was adversely affecting crops and land in the Mexicali Valley in Mexico. The problem resulted from the Welton-Mohawk Project in the United States, which brought irrigation water to the Imperial Valley in California but pumped saline water back into the Colorado. The issue received much publicity and attention in the late 1960s and early 1970s, and in 1973, the United States decided to construct a desalination plant near Yuma, Arizona. The plant is still a few years away from an operational status. In the meantime, the United States has taken other steps to reduce the salinity level and is currently meeting agreed joint standards without the plant. While this problem was very much in vogue in the early 1970s as a troublesome issue, it is generally agreed now that it is less important than air and water pollutants in urban areas along the border.

Ocean oil spills constitute another major transboundary environmental hazard. This was exhibited forcefully in the 1979 Ixtoc I oil well blowout in the Gulf of Mexico which affected the coast of Texas. Before the Ixtoc incident, the United States and Mexico were exploring the possibility of agreeing on some form of early warning systems for such incidents. In early 1980 the nations agreed to the Joint Marine Contingency Plan, which establishes response teams from both nations and a center to coordinate early warning and response to future oil spills or well blowouts. The plan is now operational.

The desirability of fruitful and reciprocal relationships between the United States and Mexico, important for any contiguous countries, are reinforced by Mexico's huge energy reserves. The past checkered record of the United States in its relations with Mexico contributes to an atmosphere of mistrust, and progress in cooperation in any policy sector can be affected. Despite this, in the past several years the level of awareness and the level of cooperation in dealing with transboundary environmental problems has increased. The population along the border is increasing rather dramatically, particularly in the El Paso-Cuidad Juarez and the San Diego-Tijuana areas. Transboundary issues are likely to become more numerous and serious. With the advent of the Reagan administration, the U.S.-Mexico Consulting Mechanism, with functional working groups, was disbanded. The governments do carry out consultation through a joint trade commission which emphasizes the important economic and commercial relations between the nations. Border problems under the Reagan administration have been treated through a pragmatic approach, and a temporary Border Relations Action Group was established. The extensive interagency contact and activity continues under this new administrative mechanism. The past experience in mutual efforts and the working relationships that are in place should enhance the possibility of amicable solutions to future transboundary environmental problems on our southern border.

Transboundary issues - the United States and Canada

The history of good relations between the United States and Canada has fostered a close and cooperative relationship. Few bilateral relationships have been so free of animosity, suspicion, hostility and war. Despite these advantages, relations in recent years have required more care and attention. One excellent example is the East Coast Fishing Treaty.

This treaty, once described by the Canadian Secretary of State for External Affairs as the most serious dispute between the United States and Canada, was approved by both governments during the Carter administration and approved by the Canadian Parliament. The treaty was complicated by a boundary dispute in the Gulf of Maine area. Ratification of the treaty by the U.S. Senate was held up because of the opposition of senators from the New England states. The Reagan administration withdrew the fishing treaty and concentrated on having the treaty deal with the boundary dispute which would provide a basis for the settlements of the fishing water dispute. Early in 1985, a decision by the World Court, set boundaries that New England fishermen felt benefited Canadian fishermen disproportionately. And by later in 1985 the fishing industry in New England was hoping to get the boundaries changed through further legal action.

This is but one of the transboundary environmental issues that characterize the U.S.-Canadian border. Fortunately, the United States and Canada have a history of mutual cooperation in regard to at least some transboundary problems. In 1909 the countries entered into the Border Waters Treaty. The treaty, which was amended in 1964, establishes guidelines and an organizational instrument for dealing with problems associated with the rivers, lakes and other bodies of water along the 5,500-mile frontier. It established the International Joint Commission, a binational organization, to deal with specific boundary water problems at the request of the two governments. Since the treaty has been in force, the IJC has been referred over 100 international projects dealing with a variety of boundary or transboundary issues.

The International Joint Commission has three principal functions:

(1) The exercise of quasi-judicial powers in approving or withholding approval of applications for the use, obstruction or diversion of boundary waters on either side of the border that would affect the natural level or flow on the other side. This function is the commission's most authoritative. Applications are made to the commission by the private firm or authority planning or constructing the project. If the commission approves, the conditions are binding on both nations and private parties;

(2) A second function is the undertaking of investigations and studies of problems at the request of either or both governments. These requests are known as references. The commission reports its findings to the national authorities and makes recommendations of specific actions. If requested, the commission conducts surveillance and monitoring of implementation of approved recommendations;

(3) Another function provides that the governments may refer a question to the IJC for decision, rather than merely reporting and recommending action. (The governments have never used this method of referral.)

The commission completes its work largely through the use of international advisory boards of equal numbers of experts from each country. In 1980 there were approximately 30 such boards. Most of the members are scientists, engineers and technical personnel. The boards are empowered to conduct studies, carry out field work and perform surveillance tasks, and they usually have representatives of federal, provincial, state and local jurisdictions. The members are charged by the treaty to serve as individuals and not as representatives of their governments or agencies. In conducting their work, these advisory boards hold public hearings at which testimony from public officials, private interests, academic institutions and public interest groups are heard. The boards report conclusions and recommendations to the IJC. The agreement which has the most significant impact and which covers the largest geographic area is the Great Lakes Water Quality Agreement. This pact resulted from an IJC project in 1972. The agreement, which was revised in 1978, established water objectives, and aims at restoring and enhancing water quality in the Great Lakes.

With over 5,000 miles of shared border, there are a large number of potential trouble spots for transboundary environmental problems between the United States and Canada. They include the long-range transport of air pollution from coal-powered plants in the U.S. and Canada, which was noted by an IJC advisory board as early as 1977. Pollutants originating from emissions from coal-powered plants in the United States are transported through the atmosphere and deposited in lakes in the northeastern United States and Canada. These pollutants, known as acid precipitation or acid rain, may reduce the capacity of these bodies of water to support marine life (and affect forest growth, crop production and the reproduction of coastal as well as inland fish). In 1978 the two governments created the Canada-U.S. Research Consultation Group on the Long Range Transport of Air Pollution. It later confirmed that the long-range transport of air pollution does occur across the Canadian-U.S. border, as well as within Canada and within the United States. The research found that about one-half of the acid precipitation in Canada originated in the United States while that originating in Canada with a southward flux is substantially less than domestic U.S. emissions.

There is widespread public awareness of the issue in Canada and concern over the possibility that approximately one-half of Canada's acid rain originates in the United States. An important interest group, the Canadian Coalition on Acid Rain, represents over 1 million members. A representative of the coalition recently stated that acid precipitation is responsible for rendering over 200 lakes sterile and threatens almost 50,000 others in Ontario. Thus far the formal governmental response of the United States has been only to increase research. But the governors in New England, equally as concerned about damage to the lakes, forests, fisheries and overall air quality in their states, have joined with the premiers of the eastern Canadian provinces in an agreement to reduce the emission of sulphur dioxide within their regions, whether or not their national governments take such action.

Water quality in the Great Lakes is a transboundary issue of huge dimensions. Nonetheless, the two countries have, particularly since the first Great Lakes Water Quality Agreement in 1972, made some progress in meeting water quality objectives and decreasing concentrations of toxic substances in some areas. There has also been an increase in the number of industrial and waste-treatment facilities in compliance with domestic pollution control programs. Yet, because of the magnitude of the problem, Great Lakes water quality will continue to be a significant issue. In 1979 the Great Lakes Water Quality Board, formed under the auspices of the IJC to carry out the terms of the agreement, reported that water quality objectives established under the agreement were still being exceeded in 48 areas. In 1981, while recognizing that progress had been made, the board stressed three priorities for further emphasis: control of toxic substances, cleanup of 39 geographic areas of concern, and the development of an ecosystem strategy. More recently concern has increased over pollution of the Niagara River resulting from U.S. waste and waste-water operations, and their impact on health and property in Canada. The IJC recommended that the governments halt all discharges in the Niagara until a comprehensive study could be made of toxic substances entering Lake Ontario. Extensive monitoring of the Niagara is being carried out by the governments, and in 1982 the EPA reported that water and air quality in the Niagara frontier had "improved markedly" but that the Niagara River is still a significant source of PCBs, pesticides and other chemical substances.

Still another troublesome transboundary issue with environmental implications, the controversy over the diversion of the waters of the Missouri River in North Dakota, known as the Garrison Diversion Unit, would irrigate over 200,000 acres of land to provide municipal and industrial water support to 14 communities and to furnish recreation, fish and wildlife opportunities in North Dakota. However, a portion of the water would enter the Souris and Red Rivers in North Dakota, and return seepage and waste from municipal and industrial systems to the province of Manitoba, Canada. Construction began in 1967.

After several years of discussion and negotiation through normal diplomatic channels, the governments asked the IJC to examine the transboundary implications. The IJC found that the project would have an adverse impact on some of the important biological resources in Manitoba and would cause significant injury to health and property. The IJC recommended that construction of parts of the project be delayed. The Garrison Diversion Unit has not yet been fully constructed and remains a thorny issue between the two countries. The plan still contains elements that could have adverse effects in Manitoba. Some interest groups and individuals on both sides of the border have opposed the project, questioning both its economic value and intrinsic merit in North Dakota. Canadians are particularly concerned over the long-term impact on the Hudson Bay watershed, and Canada remains concerned that this project will be constructed according to the original plan. Interestingly enough, the shoe may shortly be on the other foot. Plans are being considered by Canada that could affect the quantity, and possibly the quality, of water flowing into Lake Champlain from the Canadian portion of its drainage area.

Conclusion

The discussion of transboundary environment issues in North America reveals that there has been an increasing awareness of such problems, a record of cooperation and considerable activity in addressing these issues in the past decade. In both the U.S.-Mexican and the U.S.-Canadian cases, new organizational forms have been developed to bring transboundary issues to a higher policy-making level. Thus the countries have had frameworks to manage transboundary relations. A major advantage for the future lies in the improved capacity to identify problems before they become excessively disruptive.

Cooperative efforts between the countries are made more complex by the differing political systems. While ultimately under national control, transboundary issues are complicated by non-parallel governmental jurisdictions. In Mexico the government is highly centralized, with no local or state autonomy. In the United States, local and state jurisdictions exercise considerable autonomy within the limits of national guidelines and standards. In Canada provincial governments have considerable powers. These variations can cause discontinuities and prolong negotiations and complicate cooperative arrangements. For example, there are two federal, 11 state, and one provincial government involved in efforts to clean up the Great Lakes, in addition to literally hundreds of smaller jurisdictions.

There are also significant institutional and bureaucratic obstacles which hamper transboundary environmental cooperation. This is true both within the national structure and in the intergovernmental organizational frameworks. The working groups and interagency activities in both relationships have representatives from several government agencies, and IJC advisory boards include representatives from national, state, provincial and local governments. In addition, non-governmental groups, especially in the United States, influence the policy-making process. This is particularly true in the case of IJC projects which require public hearings.

The two transboundary environmental policy contexts addressed are of differing perspectives. The United States and Canada are both industrialized nations with a long and friendly history of good relations and a record of cooperation in boundary issues. Because of the concentration of industry in the northeastern United States and southeastern Canada, the kinds of environmental problems caused by industrial activities are more common and more pronounced. There has been more experience in coping with pollution, both domestically and internationally. The scope of the task is simplified in a geographic sense because only two countries occupying huge land masses are involved. Sources of pollution are easier to trace. Because of these considerations, useful comparison of the U.S.-Canadian experience to Europe are limited. With 34 countries lying in a relatively small land mass, Europe is much more vulnerable to problem relationships arising from transboundary pollution. It should be stressed, however, that even with the advantages which characterize the U.S.-Canadian relationship, transboundary environmental disputes have had disruptive effects and may become more serious. This underlines the vital need and importance of devising organizational frameworks to deal with transboundary problems in more complex international settings.

The U.S.-Mexican case is also somewhat unique in that most lesser developed countries do not lie contiguous to an industrialized country. Because of this, the example is limited in its utility as a model of transboundary environmental problems and solutions between developing nations. Mexico must deal with U.S. national and other authorities who generally perceive environmental protection as an important national and international policy issue. There is reason to believe that increasing contact with the United States in environmental issues has contributed to a realization in Mexico City that Mexico can gain from the U.S. experience. Many of the U.S.-Mexican transboundary issues are "people" problems, not generated by industrial activities but from automobile traffic, open burning and sewage discharges. Alleviation of these border issues does not entail the degree of adverse impact on economic activities more prevalent in regions of extensive industrial production and measures to correct these problems are not as controversial. The border region of Mexico is some distance from the heart of the country's industrial region and the capital city. In those areas environmental problems are likely to be more severe, and Mexico's cooperation and interaction with the United States can benefit Mexico far beyond the scope of the border region.

The North American experience suggests that significant progress has been made in developing organizational forms to carry out joint cooperation and in managing transboundary environmental problems. This state of affairs should prove helpful in anticipating potentially disruptive issues before they become less manageable. The experience also suggests the usefulness of such organizations for those instances in which even in very favorable bilateral settings, transboundary issues become irritants to general relations between countries. As transboundary problems proliferate and are more disruptive to relations between nations, greater use of these organizational forms is indicated.

[Editor's Note: Part Three gives an update on the collaboration presently going on between the New England states and the Eastern Canadian provinces.]

4. Environmental Quality Management in the United States in the 1980s and Beyond

Blair T. Bower
The Conservation Foundation
Published in 1984
Volume 4,Issue 4 of
The Environmentalist

4 Only a very wise individual or a fool would have the temerity to attempt to characterize and appraise the status of and prospects for the dynamic problem of environmental quality management in an economically complex, socially and governmentally multi-faceted, geographically large country such as the United States. I certainly do not claim to be the former, and I hope I am not the latter. Nevertheless, I will attempt to carry out that task.

Two caveats are in order. First, the following remarks represent the appraisal of one individual; they do not represent the view of any U.S. governmental or non-governmental organization. Second, the purview is limited to that sector of environmental quality management generally referred to as "environmental pollution," the class of phenomena resulting from the discharge of "leftovers" of human activities into the environmental media of air and water and on the environmental medium of land. Such sectors as occupational health, urban design, housing quality are excluded, not because they are unimportant but simply to provide operational limits to the scope of the discussion.

The remarks are organized in three sections: (1) institutional context of environmental quality management; (2) accomplishments and problems in a few subsectors; and (3) prospects.

Institutional environmental quality management

In order to understand and appraise environmental quality management (EQM) in the United States, one must understand the context, in terms of the institutional milieu. The United

States is a federal system of government, with three principal levels — national, state and local — and significant powers including raising revenues — at all three levels. Each of these levels has varying types and extents of responsibilities with respect to EQM. In some regions there is an additional, regional layer of government, intrastate or interstate, such as regional water quality boards and air quality management districts. In the United States there are approximately 3,000 counties; 18,500 municipalities; and 17,000 townships.

Not only is there a multiplicity of governmental units, but there are different organizational structures for the agencies having responsibilities for EQM. Some agencies at state and local levels have only regulatory authority, i.e., granting of permits and setting conditions on (pollutant) discharges. Others can perform multiple functions such as: build and operate facilities; set charges on, as well as physical conditions for, discharges; allocate available assimilative capacity; make grants.

Four other points are critical for understanding the institutional milieu in the United States.

One: both legislative bodies, such as Congress and state legislatures, and executive agencies have responsibility for initiating legislation to establish policies for EQM, e.g., specifying general or specific goals, specifying procedures, specifying standards, making provisions for grants for construction, specifying cost-sharing formulae, specifying tax benefits for "pollution control" activities. The legislative bodies also have "oversight" responsibility, i.e., responsibility for checking to determine if, and how well, the executive agencies have carried out the legislation.

Two: state and federal courts have important roles to perform in EQM. A very critical responsibility of the court at the highest level, the Supreme Court, is the determination of the extent to which, and under what conditions, federal law can preempt state law, unless expressly forbidden by an act of Congress. This issue is of major importance, for example, with respect to the disposal of nuclear wastes. But the courts far more often provide a mechanism for raising issues with respect to decisions

— or not making decisions — by executive agencies, such as failure to meet deadlines for setting discharge standards, failure to produce adequate environmental impact statements, failure to enforce regulations.

Three: in all states there is an agency — generally termed the public utility commission — with responsibility for reviewing and modifying or approving rate schedules proposed by utilities, primarily electric utilities but occasionally also water and sewer agencies. In carrying out this responsibility, these agencies can have major impacts on environmental quality management — for example, by requiring the utilities to carry out energy conservation programs, by requiring time-of-day and seasonal pricing, and by careful review of submitted cost estimates.

Four: there are non-governmental environmental groups which are significant participants in EQM at all levels. The activities of these groups include: data collection and analysis; review of agency EQM proposals; preparing alternative EQM strategies; lobbying legislative bodies; testifying at hearings; proposing legislation; bringing court suits. These groups represent a critical, essential part of the institutional arrangement for EQM in the United States.

Accomplishments and problems

To be able to make any statement concerning prospects for maintaining or improving ambient environmental quality over the next decade or so requires both assessing past performance and identifying problems associated therewith. Only a few subsectors will be cited herein, but these are sufficient to emphasize both some specific EQM problems and some generic EQM problems, the latter relating to problems which occur in most or all subsectors.

No attempt is made to develop rigorous, unambiguous definitions of subsectors. One can classify by environmental medium, by type of material discharged, by geographic scope. However, overlapping is inevitable.

Water quality

Considering the decade of the 1970s and water quality indicators reflecting the so-called conventional "pollutants," — dissolved oxygen, suspended solids, bacteria, dissolved solids, phosphorus (to reflect nutrients) — the policies and programs of the three levels of government have resulted in improved ambient water quality in some areas, degradation in some areas, and basically no change in other areas. Given increases in population and output, particularly in some regions, it is not an insignificant accomplishment to have done even that well. Some of the worst pollution problems in the United States have been reduced: e.g., salmon have reappeared in the Connecticut and Penobscot Rivers in New England; the catch of shad has increased significantly in the Delaware River; eutrophication in Lakes Ontario and Erie has been slowed; ecological productivity of some estuaries (for example, several in Florida) has increased as a result of reduced discharges, although in other bays (such as the Chesapeake) and estuaries productivity has decreased.

Major problems are: maintaining continuing compliance by dischargers to surface-water bodies with discharge (effluent) standards and other conditions imposed on the dischargers with respect to conventional pollutants; groundwater contamination; and managing toxic pollutants. Simply to grant a permit to discharge under specified conditions according to specified standards is only the first step in achieving compliance. Monitoring, inspection and significant sanctions for non-compliance are essential. The picture is mixed. Considerable day-to-day compliance has been achieved by many private sector activities, but the record for public sector activities is not as good. Studies have shown that the number of major municipal sewage treatment plants not meeting their discharge standards from one-fourth to one-half the time is significant. An even greater problem involves installations of the national government.

About one-fifth of total water withdrawal in the United States is from groundwater, and about half of the U.S. population obtains its domestic water supply from groundwater. However, what is critical is the increasing number of discoveries of contamination of groundwater sources/aquifers. Hundreds of wells have been closed because of contamination from toxic organic chemicals and reports of new closures are at least weekly occurrences. It must be recognized that when the contamination first occurred is often not known. Only within the last decade or so has cognizance of the problem or potential problem increased so that tests for water quality are more often made. The publicity given to various incidents has increased awareness. Typically, contamination of groundwater aquifers is irreversible.

In the United States about 60,000 chemicals currently are manufactured, imported and marketed. Of these, only a small proportion has been tested for long-term effects on the environment, on human health and on other species. One hundred and twenty-nine "priority pollutants" and 65 "priority toxic pollutants" have been identified under the Federal Water Pollution Control Act Amendments of 1972. However, although estimates have been made of the quantities of "toxic chemicals" generated by major industrial categories, relatively little is known about the types and quantities generated by individual unit processes and unit operations at individual activities, and even less is known about the disposition of these chemicals into the environment. Until recently there were no requirements that these materials be measured in discharges. But it should be emphasized that various toxic chemicals are discharged not only from industrial activities but from municipal sewage treatment plants, in urban storm runoff, from mine tailings, in runoff from agricultural activities, and from municipal and industrial landfills, which — more often than not — are simply dumps.

Soil erosion

Despite governmental programs to reduce soil erosion from agricultural lands initiated in the latter half of the 1930s and still continuing, soil erosion continues to be a major problem, with adverse effects both on-site and off-site. In some areas of the United States, certain policies in the last decade may actually have exacerbated the problem. The rate of soil erosion on cropland in 1977 ranged from about five metric tons/hectare/year on the best land to about 120 metric tons/hectare/year on the poorest. Soil erosion on about 30 percent of agricultural land was less than about two metric tons/hectare/year, accounting for about 2 percent of total erosion from agricultural land. Soil erosion on only about 3 percent of agricultural land was more than 56 metric tons/hectare/year, accounting for almost 30 percent of total erosion. To put it another way, in 1977, for each ton of corn raised by Iowa farmers, an average of five tons of soil eroded from the land on which the corn was grown.

Multiple factors contribute to the continuation of soil erosion from agricultural land as a significant environmental quality problem. One is the uncertainty about the impact of soil erosion on productivity. Relatively little quantitative information exists. Effects of erosion on productivity — and there is evidence of significant adverse effects for some soils — are masked by the increase in average crop productivity nationally as a result of more fertilizer and pesticide inputs per hectare and better varieties. Additional inputs can compensate for soil erosion and loss in natural productivity only up to some level, for most soils.

Another factor is the short time-horizon of decision-making by many farmers. Loss in productivity may well not be significant, in terms of reduced profits, over a generation. Therefore, investments to reduce soil erosion are not considered justified. This may be even more true for farmers who lease rather than own the land they farm.

There are at least 30 different programs dealing with soil conservation in the U.S. Department of Agriculture. These involve, for example, financial assistance and technical assistance. However, a Department of Agriculture study found that over half of the activities were undertaken on lands eroding at a rate within the tolerable range, whereas only about one-fifth were on lands eroding at more than 13 metric tons/hectare/year. The latter accounted for 85-90 percent of the total erosion.

Finally, within the Department of Agriculture there have been conflicting goals. The push for production of agricultural products for export, and the provision of price supports has led to such practices as: more intensive farming in terms of inputs per hectare; plowing to the fence line; removal of terraces and grassed waterways; plowing straight instead of on the contour. At the same time, other agencies of the department have been pushing programs to reduce erosion by measures opposite to the above. Recognition has increased that these inconsistencies should be eliminated and a trial program is underway in which various price support payments will not be made unless the farmer has installed and is operating various soil conservation measures.

Air quality

Over the last 10 years, ambient air quality in a considerable number of metropolitan areas in the United States has improved, as measured in terms of the concentrations of the five so-called "criteria pollutants": total suspended particulates, sulphur dioxide, ozone, carbon monoxide, nitrogen dioxide. This has been accomplished by: improved installation, inspection and maintenance of discharge-reduction devices in mobile sources (automobiles, etc); modified production processes, changes in raw material inputs, installation of discharge-reduction devices at stationary sources; and — in some cases — the closing of obsolete plants and equipment at plants, perhaps most notably in the steel industry. However, two points are clear. One: many areas are not likely to attain health-based ambient environmental quality standards with respect to the above five indicators (the "primary" standards) for some time. Two: where those standards are currently being met, they will not be maintained unless continuing compliance is achieved by both stationary and mobile sources.Achieving continuing compliance is a major problem, even more so with respect to discharges of air pollutants than pollutants in water, a conclusion reached by studies by the U.S. General Accounting Office as well as others.

With respect to mobile sources, inspections and maintenance programs for vehicles, initiated under the Clean Air Act, have clearly had positive results where they have been effectively executed. However, there appear to be pressures: to relax some vehicle emission standards; not to promulgate emission standards for certain vehicle types not yet covered; and not to initiate inspection and maintenance programs in some regions where they clearly are needed. Further, it is not clear that the recent trend of a decrease in average rate of emission per kilometer driven can continue to offset the increasing numbers of miles driven. In no metropolitan areas in the United States have transportation plans focused on reducing automobile traffic and increasing use of public transit had a significant impact in terms of reducing vehicle miles traveled. In fact, the expansion of the number of parking spaces in the central business districts of metropolitan areas accompanying urban redevelopment and the improvement of arterial highways has induced greater use of private automobiles.

As difficult as are the above-mentioned problems, at least three considerably more difficult problems associated with air pollution have become recognized in the last decade. These are: (1) fine particulates; (2) toxic materials; and (3) acid precipitation. Current regulations with respect to particulates do not differentiate among particles of different sizes. However, based on recent research, smaller-sized particulates, particularly those having diameters of less than 2.5 microns, are a potentially greater threat to human health than are large particulates. The latter are less likely to penetrate into lungs than are the fine particles. In addition, discharges of oxides of sulphur and nitrogen can be converted in the atmosphere into fine sulphate and fine nitrate particulates. The extent of generation of fine particulates is only beginning to reach the level of understanding that is necessary for the development of standards for their regulation.

Concomitant with the recognition of the problem of fine particulates has come the recognition of the large numbers of toxic materials of which many gaseous discharges are comprised. Traditionally, all particulates have simply been "lumped" into the category, total suspended particulates, undifferentiated. In reality, these particulates are often comprised of a number of substances, many of them now recognized as toxic. For example, the combustion of fossil fuels often results in fine particulates, including such substances as lead, mercury, vanadium, manganese, polycyclic organics. About 50 chemicals and synthetic organic compounds have been identified for priority consideration as hazardous pollutants in gaseous discharges. By 1982 emission standards had been promulgated for only four of these.

Recognition of the existence of acid precipitation and of some of the apparent consequences thereof was spawned in the early 1970s, particularly in Europe. As has happened with other environmental pollution problems, such as DDT, and nitrates in groundwater, once the problem was recognized in one or a few regions, efforts were directed toward analyzing situations in other regions. Acid precipitation and the increasing acidification of lakes have been found to be much more widespread in the United States than had been thought. The areas most seriously affected are the Adirondack Mountains in New York and the northeastern United States. roughly from the North Carolina-Virginia border to Maine. Southeastern Canada also is a region seriously affected; thus fact continues to be a significant diplomatic problem between the United States and Canada. Within the last few years, acidification has been found in lakes in the Sierra Nevada mountains in California.

Acid precipitation has added to the recognition of the long-range transport problems with respect to gaseous residuals. It has now clearly been shown that discharges of various materials in gaseous waste streams can be, and are, transported long distances in the atmosphere. This is not limited to materials which result in acid precipitation; there is no other explanation for the finding of DDT in Antarctica in the late 1960s and of mirex — used only in the "Deep South" in the United States — in the Great Lakes in the 1970s. Further, a study of phosphorus inputs into the Great Lakes concluded that about 25 percent of these inputs into Lake Huron and Lake Michigan and about 40 percent into Lake Superior were directly from the atmosphere, and from sources outside of the Great Lakes Basin.

Hazardous Wastes

"Probably the most important, controversial, and difficult environmental problem to emerge over the past few years has been what to do with hazardous wastes — those that are being generated, those that will be generated, and those that already are sitting in tens of thousands of sites around the country." [Source: The Conservation Foundation, State of the Environment, *1982.]*

Toxic chemicals seeping from Love Canal in New York; dioxin throughout the town of Times Beach in Missouri; kepone in Hopewell, Virginia, and the adjacent James River: these incidents resulting in national headlines represent a minuscule portion of the problem of managing of hazardous materials. That the problems represented by these materials are widespread and pervasive has become increasingly clear, as indicated by the comments made above, for example, with respect to contamination of wells and gaseous discharges. But the amounts of information required to define the problems, to develop strategies for dealing with the problems, and to establish priorities is grossly inadequate. Much less is known about the origins of these materials, the current and past dispositions thereof, and methods for handling and disposing of them than is the case for so-called "conventional" pollutants. At present there is substantial uncertainty about the number of firms involved in one or more of generation, transport, storage, treatment and disposal of hazardous wastes.

The total amount of hazardous materials being generated in the United States probably is increasing. However, the rate of increase and the locations of generation are not known accurately. The largest generator of these materials, the chemical and allied products industry, grew significantly faster in the 1970s than the average growth rate of the remainder of U.S. industry. In this case the increase in generation does not mean necessarily an increase in discharge to the environmental media. The industry as a whole has made substantial efforts to cope with the problem. For example, a survey in 1981 and

1982 of 70 chemical companies involving 535 plants found that these companies, as a whole, recycled, reused, reclaimed about one ton of hazardous wastes for every two tons disposed.

About three-fourths of the hazardous wastes generated by industrial activities are disposed of on-site, primarily on land. However, no one knows in how many locations hazardous wastes have been dumped in the past, and are being dumped at present. The Environmental Protection Agency has identified about 11,000 active sites and 9,000 to 11,000 inactive sites. Most of the active sites are owned and operated by waste generators. How many of these sites pose serious risks is not known. One study estimated that anywhere from 5-95 percent of the active sites could pose a significant threat to public health and/or to the environment.

With respect to generator-owned disposal sites, the nature and extent of environmental protection measures at these sites is largely unknown. One study of impoundments used for disposal sites found that "over 70 percent of the industrial impoundments are unlined, potentially allowing contaminants to infiltrate unimpeded into the subsurface" and "nearly 95 percent of the sites are virtually unmonitored as to possible groundwater contamination."

Even more disconcerting are the numerous incidents of illegal dumping, often termed "midnight dumping," although the activity occurs at any time of day, and any day of the year. Some of the dumping or disposal — for example, application of used oil on rural roads to reduce dust — is innocent in that the dumper does not know that the material contains toxic substances. Other illegal dumping is deliberate and has had very adverse consequences in many known cases and in many still being located. Probably the extreme of the problem is reflected in the following description of an investigation in northern New Jersey in the late 1970s.

The investigators found haulers running rampant, dumping drums of toxic chemicals under the Pulaski skyway (a major highway), in uncultivated fields, and in vacant urban lots; throwing drums off the back of moving trucks onto the shoulders of main thoroughfares; digging up streets in residential neighborhoods and filling the holes with drums; stacking countless drums in abandoned waterfront piers; abandoning stolen trucks filled with waste-containing drums in the middle of city streets; pumping tanker trucks full of chemical wastes into residential streets, creeks, municipal sewers, old water wells, Newark Bay, and the Arthur Kill, a narrow body of water between Elizabeth, New Jersey, and Staten Island, New York; mixing toxic flammable and corrosive wastes with garbage and hauling the wastes off to municipal landfills; filling the basements of abandoned inner-city tenements with hazardous wastes, then demolishing the structures; and creating large artificial lakes of toxic wastes in the Pine Barrens, the site of New Jersey's future drinking water supply." [Source: Epstein, Brown and Pope, *Hazardous Waste In America,* Sierra Club Books, 1982.]

A special fund, the so-called "Superfund," has been established to provide funds for "cleanup" of inactive sites. Even if the fund were sufficient to cover costs likely to be incurred at all sites, it would be impossible to work on all sites simultaneously; therefore, priorities must be established. At least the following criteria are appropriate for that purpose: (1) quantity of hazardous materials in the given sites; (2) relative effects of the materials on receptors, e.g., danger to human health; (3) location in relation to centers of population; and (4) location in relation to groundwater aquifers from which potable water is obtained. However, any effort to assess identified sites in relation to these criteria in order to establish priorities is itself difficult, costly and time consuming. The problem with respect to currently operational disposal sites is worse in one respect; namely, there is no special response fund for use in improving performance at these facilities.

Precursors and Prospects

Past and present behaviors of individuals and societies are often precursors of future behavior. The foregoing observations provide some bases: (a) for identifying some of the characteristics of the U.S. approaches to EQM; and (b) for appraising prospects. They reflect reactions to, attitudes toward, understanding — or lack of understanding — of environmental pollution problems and factors affecting those problems.

• Looking back over the last two decades, an observer might well characterize "the record" as reflecting movement "from one crisis to another," in terms of type of problem and geographic locale. DDT in birds; nitrates in groundwater; radioactivity in construction materials manufactured from uranium mine tailings; mercury in Great Lakes fish; the Cayahoga River on fire; kepone in Virginia; an exploding, burning landfill in New Jersey: the record of "sudden surprises" seems unending. As soon as understanding begins to be achieved with respect to one problem the next one(s) explode(s). As T. Tuchenberg, one cogent observer of the problem in the Great Lakes put it in the November 1983 issue of *The Conservation Foundation Letter*: "It seems that every year in the past decade, public attention around the Great Lakes has been captured by a 'new' chemical pollutant."

• The interrelationships among the three environmental media have all too often been ignored. Measures adopted to reduce one type of environmental pollution problem have resulted in exacerbating other environmental quality management problems or in simply transferring a problem from one medium to another. Imposing more stringent restrictions on discharges to one medium will usually increase the quantities of residuals discharged to other media. Thus, tightening standards for waste-water treatment results in the generation of increasing quantities of sludge, which requires disposal. Sludge disposal has become a major problem, especially in metropolitan areas, with no long-run solutions yet available in many cases. Tightening the standards for discharges of, for example, particulates and oxides of sulphur to the atmosphere results in large quantities of dry fly ash or wet scrubber residuals requiring disposal. The only way to reduce the problem is to reduce generation. Despite the fact that intermedia interactions are likely to increase in the future in scope and intensity, the institutional structure needed to cope with these interactions/ interrelationships is virtually non-existent at any level of government.

• All too widespread and commonly accepted are the assumptions that activities — private and public — will adhere to comply with restrictions on their discharges, i.e., 100 percent continuing compliance, and that production activities involving use of materials which can have adverse consequences if discharged will perfectly execute the use of those materials so that no discharge takes place. Both assumptions have been proven to be naive, again and again. The nature of the problem of continuing compliance was mentioned previously. What has become increasingly clear is that some of the major problems with respect to continuing compliance involve public sector activities. In an August 17, 1983, article in *The Washington Post,* C. Peterson stated that " The biggest violator of federal pollution laws may be the federal government" particularly with respect to toxic wastes. It seems to be inherently difficult for one public agency effectively to regulate another public agency/activity at the same level of government.

The problem with respect to perfect execution in production is aptly expressed in the following:

• The EPA's pesticide regulations and the practices of farmers somehow assume that complex, persistent organic molecules can be carefully deposited on one part of a farmer's acreage at a particular point in the growing cycle and yet not become incorporated in human food chains. This attitude overlooks a wide range of phenomena, such as wind-induced drift, soil residues, runoff into streams, mis-timed applications, deliberate violations of regulations, and mis-labelling errors. All of these ensure that a significant fraction of the total volume of pesticides applied in this country ends up being ingested by human beings. [Source: *Conservation Tillage*, a special issue of the *Journal of Soil and Water Conservation*, M. Schnepf, editor. 1983.]

• Decisions in other non-EQM sectors of the economy, e.g., in response to changes in factor prices, have often exacerbated environmental pollution problems.

Growing efforts with respect to energy conservation, in response to the sharp price increases in 1974 and additional increases in subsequent years, have been undertaken in homes and offices. By increasing the amount of insulation in and reducing air loss from buildings, the amount of air coming in from the outside is reduced. Where the outdoor air is cleaner, air quality inside decreases with the reduction of inflow. In addition, some of the materials used for insulation are themselves sources of pollutants. One response in the agricultural sector to increased energy costs was to shift toward, or to, minimum tillage agriculture. This shift had a positive effect with respect to one aspect of environmental pollution, by generally reducing soil erosion. However, the shift to minimum tillage required substantial increase in use of herbicides, thereby increasing pesticide discharges.

One major factor in increasing productivity in U.S. agriculture since the Second World War has been the development of various high-yielding crop varieties. These developments have led in turn to a significant increase in area under monoculture (devoted exclusively to one crop). This has at least two implications with respect to environmental pollution. One, to achieve the high yields associated with the high-yielding strains requires additional associated inputs, e.g., fertilizers, water, pesticides. Two, these in turn typically lead to increased discharges to the environment.

The push for "energy independence" in the United States, beginning in the mid-1970s, will — if continued without constraints — exacerbate environmental pollution problems. Production of oil from oil-shale plants and coal liquefaction have been proposed as major building blocks of this independence. The former is expected to result in generation of radionucleides, polycyclic organic matter, arsenic, cadmium, mercury, beryllium, selenium; coal liquefaction results in generation of polycyclic aromatic hydrocarbons, carbonyl sulphide and phenols. In addition to gaseous discharges, there are major quantities of solid residuals requiring disposal, in the case of oil from oil shale about 20 tons for every ton of oil produced.

Even tax policies developed by the U.S. Internal Revenue Service can, and do, exacerbate environmental pollution problems. For example, various provisions for depletion allowances and expensing provisions in relation to primary materials, and leasing of public mineral claims at prices substantially below market values, induce more use of primary materials at the expense of secondary materials which do not "benefit" from such provisions and whose costs are therefore relatively higher. Use of primary materials generally results in substantially more discharges to the environmental media than use of secondary materials.

• EQM is undertaken in an environment of uncertainty with respect to the effects of discharges of substances into the environment, and the effects of exposures to the resulting ambient concentrations. The extent of bioaccumulation, long latency periods, extent of synergistic effects are unknowns in many cases, for both human and other species. For some substances the ability to measure their concentrations in the environment has been developed only in the last decade. Each time the capacity to measure becomes more detailed, one or more "new" substances are discovered to be in the environment. Which substances by which paths/modes of impact comprise the most significant components of "total body burden" remains uncharted.

• The decreasing quality of raw materials, such as iron ore and copper ore, means that the quantity of residuals generated per unit of product output will continue to increase. Increased reuse of metals would slow this trend, e.g., production of aluminum ingots from recycled beverage containers requires about 10 percent of the energy required to produce the containers from raw aluminum ore.

• The problem of where to dispose of nuclear wastes has not been solved. The search goes on, with continual confrontations between one or more states and the national government. Even if the process of selecting and constructing the first high-level repository were to continue on schedule, the first repository would not be ready for use until the late 1990s. Whether or not the temporary storage sites now in use or subsequently developed will be sufficient until that time is not clear.

• The reported loss of jobs and the claimed high costs of improving ambient environmental quality, continue to be arguments raised in the political arena of environmental pollution policy and programs. This is in spite of the facts that: (1) EQM programs have created many jobs, both directly and indirectly — the latter for example in production of discharge reduction equipment; (2) there are widespread benefits from EQM programs; and (3) the real resource costs are often much less than originally anticipated, provided the incentive structure induces the search for and selection of least-cost options, particularly reduction in waste generation. The "Pollution Prevention Pays" program of the 3M Company (Minnesota Manufacturing and Mining) is an excellent example. That program, in operation since 1976, stresses the elimination of pollution at the source, i.e., the redesign or original design of production processes that will minimize the necessity for residuals disposal.

• The federal government recently has pursued a policy of decentralization of responsibility for carrying out EQM programs; however, this has been done essentially without increasing the resources available to the state and local governments. State governments often cannot compete with either the national government or with the private sector for personnel. Without personnel and supporting resources, it is clear that many state governments simply do not have, and are not likely to obtain, the capability to do the requisite tasks of EQM. Of course, states vary substantially in both resources and commitment. For example, as part of a national report prepared in 1982 and relating to environmental quality activities, The Conservation Foundation evaluated state efforts to maintain environmental quality. Factors considered included the voting records of the state's Congressional delegation, state regulatory programs to protect the environment in the state, state expenditures on environmental quality and 20 other factors. Minnesota had the best rating, followed closely by California, New

Jersey and Massachusetts. Given the variability among the states in capability, one would expect the national government to exert some leverage in order to insure some minimal level of execution by the states. However, just as it seems very difficult for one federal agency to regulate another federal agency, there are few sanctions which the federal government could, or has been willing to, impose on states to induce actions.

• Lest the impression that the behaviors in the United States over the last decade yield only "negative" prospects, the following evidence should be considered:

— Support for maintaining strict pollution control laws remains high among all groups of the U.S. population, despite depressed conditions in some economic sectors and relatively high unemployment. At least part of this response stems from the continual disclosures of various surface and groundwater bodies and land surfaces contaminated with toxic materials; no section of the country has "escaped." Public opinion polls continue to find large majorities, 60-80 percent of respondents, in favor of retention of strong controls and even larger majorities rating various air and water pollution problems as critical problems.

— Membership and financial resources of environmental groups, such as the Sierra Club, Environmental Defense Fund, Natural Resources Defense Council, have increased very substantially in the past few years. Activities of these organizations include not only education, monitoring of legislative activities and proposing legislation, but significant efforts to aid enforcement. For example, the Natural Resources Defense Council is filing lawsuits against major private industrial activities in New York and Connecticut for violations of their permits to discharge to water bodies.

— Energy use has become more efficient in the major end-use sectors of the economy, thereby reducing the demand on environmental resources. Between 1973 and 1981, real gross national product (GNP) grew by about 20 percent while energy use actually decreased slightly. In 1981 energy use was 2.5 percent below that in 1980, while real GNP rose 2 percent. Of course the major stimulus was the sharp increase in energy prices beginning in 1974. However, the rapidly increasing costs of new power plants has stimulated the adoption of energy conservation programs even among private utilities, given that such programs are by far the least-cost approach to equating demand and supply.

— The last decade shows other indications. By 1984 ten states were requiring deposits on non-returnable containers. Fifty percent of all aluminum cans were recycled in 1982. There are more bicycle trails and bicycle commuters. One finds more unbleached paper towels in commercial facilities. There is a sufficiently large demand for organically grown foods and foods without multiple additives to support a small but viable "natural foods sector."

— The decline of the "smokestack" industries has resulted in the deactivation of some of the most polluting industrial facilities. Where particular outputs of these activities are still desired, new investment tends to be in less-polluting — as well as productively more efficient — technology.

Part Three

"Mistakes are part of learning. There is nothing good or bad about behaviors or perceptions that do not work, they simply have to be given up and replaced by behaviors and perceptions that DO work."

R. Buckminster Fuller

Putting the Pieces Together

Bucky Fuller's quote reminds us of the importance of searching out "behaviors and perceptions" that will work in our changing world. Although many such initiatives have appeared in the preceding pages, this final section of the book will focus on still more responses that national and global environmental problems have stimulated — responses that are important both in terms of their potential for resolving serious environmental problems, and for the clues that each may offer regarding the kinds of initiatives that other forms of global interdependence may require us to consider.

The section will start off with an eclipsed list of events, publications and institutional initiatives that have occurred within the recent past, and then explore some specific examples in greater detail. We will look first at several serious environmental issues which have moved well into the public and political consciousness and for which strong management mechanisms are evolving. We will then examine other environmental issues, whose seriousness is as great, but for which management mechanisms are just beginning to emerge, either because of their political or scientific complexity or the fact that they have come only more recently into the consciousness of major leaders. Contacts for each of the cited efforts appear at the end of the book for those readers who wish to learn more about them.

1. Timeline of Responses Stimulated by Global Environmental Issues

1948 - *International Union for the Conservation of Nature and Natural Resources* was established by people who were disturbed by the signs of increased encroachment by human settlements and activities on habitats of key wildlife species and natural systems. IUCN was set up as a hybrid organization — creating a necessary bridge between governmental, scientific and non-governmental interests — to address conservation problems that crossed political boundaries. It is one of the few organizations that countries and organizations can join as equal partners. It has had, from the beginning, close cooperation between scientists and policy makers from all parts of the world. Over the years IUCN has set up seven commissions (Parks; Ecology; Policy; Law and Administration; Environmental Planning; Species Survival; Education). Representatives of member nations and organizations serve on these voluntarily. A considerable amount has been accomplished, including the publication of Red Data books that report on the status of endangered species; and meetings that have led to or contributed to the establishment of international agreements regarding protection of key species, establishment of parks, preservation of major international cultural sites, etc. An early group to recognize the close relationships between the needs to protect natural systems while meeting the needs of human populations, IUCN did much to shape the concept that has evolved into "sustainable development."

1961 - *World Wildlife Fund International* was established by many of the same individuals involved in setting up IUCN. Originally perceived as a way to raise money to fund the scientific projects and activities of IUCN, WWF has become a powerful entity on the international conservation scene in its own right. Most of its project support funds go directly into preservation of individual species. WWF/International and national offices have done a great deal to raise broad public awareness about endangered species and other international conservation issues. WWF/U.S. has recently merged with The Conservation Foundation of Washington, D.C., one of this country's most respected conservation organizations, which has, in recent times, focused considerable effort on developing improved mechanisms for resolving environmental conflicts.

1968 - *The Man and the Biosphere (MAB)* Program at *UNESCO* was initiated by the United Nations General Assembly. Designed to specifically look at man-disturbed environments rather than pristine natural environments, MAB was to explore both positive and negative types of human impact in ways that could help design productive ways for human intervention in natural systems to occur. Teams of scientists and experts from many countries have collaborated on projects, building a communications network that has been highly valuable. The U.S./ MAB program had been particularly strong; however, it is likely to be affected negatively by the recent United States withdrawal from UNESCO.

1972 - In June 1972 delegates and advisors from 113 nations gathered in Stockholm, Sweden, at the *United Nations Conference on the Human Environment.* These global leaders came together in recognition of the need to discuss rising environmental problems all around the world. Excessive and mismanaged resource use was causing pollution-fouled air and water in heavily industrialized areas, while the very struggle for survival — for adequate food, water, firewood and shelter — was threatening resource bases in rapidly developing sectors of the world. There was growing awareness that the fates of all countries and all people — young or old, wealthy or poor — were inextricably intertwined. Ironically, the advent of nuclear weapons had helped build this new consciousness. Nuclear fallout, detected by analyses of air, water and other resources far from the test sites, had triggered an awakening realization of the infinite interrelationships among Earth's life-support systems.

Four years of preparatory work, led by the secretary general of the conference, Maurice Strong, preceded the gathering — years during which background data was gathered on increasing air and water pollution in the industrialized nations, population pressures in certain areas, and fears that non-renewable resources (oil and coal in particular) were being used at too fast a rate. The meeting was attended by many of the people who were involved with IUCN and WWF, and thus some of the conservation-oriented issues were presented as well. Lesser-Developed Countries attended, but more from a defensive posture. Leaders from these nations were fearful that the cries to "reduce pollution," "slow the rate of consumption of non-renewable resources" and "limit population growth" were just more ploys to keep their people in a state of repression. Over time, however, these countries have largely become strong supporters of many of the basic thrusts of Stockholm. Their well-articulated opinions that environmental protection could not succeed without development — that poverty was a major source of environmental degradation — helped shape the concept of sustainable development and of the other ideas that were to emerge in 1980 in the World Conservation Strategy.

In December of 1972, the United Nations General Assembly, acting on the recommendation of the Stockholm conference, established the *United Nations Environment Program* to provide catalytic coordination among the existing organizations with responsibilities for activities that affected the global environment. These included *UNESCO, United Nations Development Program, World Health Organization, Food and Agriculture Organization of the United Nations, World Bank* and others. In addition to serving as the focal point for environmental action and coordination within and even beyond the U.N. system, UNEP was instructed to develop an international surveillance network, to monitor current conditions and changes in the environment, and serve as a global early-warning system; and to advocate sustainable development — encouraging and maintaining the link between environmental principles and practices and the economic and social well-being of the human population.

UNEP was the first U.N. organization to be headquartered in a developing country (Kenya). The reduced contacts and services (compared to organizations located in Paris or Geneva), the "soft" mandate of the agency, and its minimal size (a worldwide professional staff of about 170 and an annual budget of $30 million) have been difficult handicaps to overcome. Nonetheless, in activities like its Regional Seas Program, UNEP has been able to define areas for cooperation between member nations characterized by great differences in political regimes, size, natural resource base and levels of development. And its expanding global monitoring program, described in more detail later in this chapter, has the potential for providing much highly useful data to planners and decision-makers all over the world.

1974 - *The International Environmental Education Program,* supported by UNESCO and UNEP, was set up in response to Stockholm recommendation #96, which recognized the key role of education in achieving the other goals of the conference. Under a several-year program, the IEEP surveyed existing environmental education programs and needs all over the world (published by UNESCO in 1977 as *Trends in Environmental Education*) and conducted a series of meetings that culminated in the UNESCO/UNEP sponsored meeting in Tbilisi, U.S.S.R. in 1977, when top education ministers from many nations adopted global environmental education goals. The IEEP has continued to support small pilot projects that have good potential for replicability. Seventy-five percent of the UNESCO member countries have benefited directly from the IEEP.

1970s-to-mid-1980s in general — Conditions like the energy crisis of the mid-seventies, combined with reports like the Club of Rome's *Limits to Growth,* the Brandt Commission Report, *North-South: A Program for Survival,* the World Bank's *1980 World Development Report,* and the 1980 *Global 2000* Report to the

President (Carter) and its counterpoint *Our Resourceful Earth* preliminary findings concerning the potential impacts on global environmental systems of nuclear war, and, most recently, the African famine have brought the issues of global environmental interconnectedness and interdependence to broad awareness. The result has been increased attention by governments, industry, foundations and non-governmental organizations. Some examples are:

1973 - Establishment of *The Institute of Ecotechnics,* in London, for the purpose of beginning and developing a new discipline interrelating man, his culture and his techniques with the totality of planet Earth. Today the institute conducts cost-effective demonstration projects concentrating on ecological risk and opportunity. Present projects are placing emphasis upon rainforests, savannas, deserts, city centers, rivers and coral reefs.

1974 - Establishment of the *WORLDWATCH INSTITUTE*, in Washington, D.C., under initial support by the Rockefeller Brothers Fund. Worldwatch, a leading center for information and analysis, studies major trends and new activities that are affecting the planet. It has prepared numerous (nearly 70) excellent short and inexpensive reports on individual aspects of global environmental issues, and, in 1984, began the publication of annual *State of the World* composite reports.

1980 - Publication by IUCN of the *World Conservation Strategy* — the first attempt to build broad support for "sustainable development," i.e., development that meets basic needs of people without undermining the resource base. The WCS was supported and assisted by WWF and UNEP, with heavy involvement by the Food and Agriculture Organization of the United Nations and other such agencies with related programs. National Strategies for implementation of the WCS are under preparation in more than 30 countries. A major conference is to be held in Canada in 1986 to assess progress and share experiences. The United States has not yet chosen to develop a strategy.

1980 - Establishment of the *Global Tomorrow Coalition* as a result of the *Global 2000* report. The coalition is an alliance of 100 organizations with combined memberships of over 6 million U.S. citizens. GTC works with and through members like the Natural Resources Defense Council to strengthen U.S. legislation on national and international environment and development issues, and disseminate information to its member organizations and the broader public through publications, workshops and conferences.

1982 - Establishment of the *World Resources Institute,* Washington, D.C., under funding from the MacArthur Foundation, to study and develop policy guidelines regarding large-scale resource, environmental and population problems viewed from a global perspective. In 1983 a conference of leading thinkers led to the publication of *The Global Possible: Resources, Development and the New Century.* The Institute has ongoing policy studies regarding carbon dioxide, toxic wastes, genetic diversity, multinational corporations, and the U.S. stake in the global environment. [Results of a recent effort to develop management techniques for tropical rain forests are discussed more thoroughly later in this section.]

1983 - Modifications to the *World Conservation Strategy* to incorporate human population problems and needs and protection of cultural traditions into the development and implementation of national strategies. The former change was accomplished through collaboration between IUCN's Ecology Commission and the International Planned Parenthood Federation.

1983 - Establishment, by the U.N. General Assembly, upon the recommendation of the UNEP Governing Council, of the *World Commission on Environment and Development* — a team of 23 wise men and women charged with the awesome task of "elaborating the [global] environmental perspective to the Year 2000 and Beyond" and directed to propose long-term environmental strategies for achieving sustainable development. The recommendations are expected to include proposals for modified or new types of legal and institutional approaches, and innovative approaches to funding. The commission, scheduled to submit its report, *Common Future* to the U.N. General Assembly in 1987, includes representatives from the United States and Canada — the Honary William D. Ruckelshaus, former Administrator of the U.S. Environmental Protection Agency and Mr. Maurice Strong, whose paper begins this book. Other representatives on the commission come from Algeria, Brazil, People's Republic of China, Colombia, the Federal Republic of Germany, Guyana, Hungary, India, Indonesia, Japan, Mexico, Nigeria, Norway, Saudi Arabia, the Sudan, Yugoslavia and Zimbabwe — giving the full commission quite a remarkable range of viewpoints on global environment and development issues. The WCED is chaired by Ms. Gro Harlem Bruntland, former Prime Minister of Norway, who noted the urgency of this effort, "as we move into the next century, building another world on top of the one we have, and doubling, at least, our demands on the planet's ecosystems."

1984 - Versailles, France, *The World Industry Conference on Environmental Management*, co-sponsored by UNEP and the International Chamber of Commerce. WICEM brought together over 500 representatives from major multinational corporations, international agencies, national governments and non-governmental organizations to define a broadened agenda for cooperation. Occurring just weeks after the poison gas tragedy at Bhopal, India, much emphasis was on the issue of protecting workers and the general public from hazardous wastes. A related follow-up program was recently announced by the U.S. Agency for International Development and the UNEP-sponsored World Environment Center in New York City. Under a pilot project to aid industrializing nations, U.S. corporations will send volunteer experts to help improve or create new systems at the factory level for responding to industrial accidents — particularly, safeguarding populations living around chemical plants, petrochemical operations and other industrial complexes. The new emergency response management project will be tested first in the Middle East. One country will ask AID officials to provide a team of experts to assess factory contingency plans to prevent and respond to industrial emergencies. The U. S. experts will work closely with local plant managers to create a system to meet emergency needs. Among the U.S. companies involved are Combustible Engineering, Dow Chemical, Tenneco, Inc., 3M and the Weyerhaeuser Co.

1984 - Publication of *The Cold and The Dark: the World After Nuclear War*, edited by Ehrlich, Sagan, Kennedy and Roberts, the proceedings of the Conference on Biological Consequences of Nuclear War. The conference focused on the results of studies carried out by hundreds of American and many Soviet scientists, among them climatologists, biologists, physicists and mathematicians — who examined the potential impacts of a nuclear exchange upon the natural systems that support all forms of life in all nations.

1984 - Publication of *GAIA: An Atlas for Planetary Manaagement*, general editor, Norman Myers. This book's wide range of subject areas and extensive use of colored graphics helps convey to the lay reader the nature of individual global environmental issues and of their important effects on our attempts at solutions to other major societal concerns. The title, GAIA, was taken from the hypothesis first put forth by James Lovelock, in *GAIA: A New Look At Life on Earth* (1979), in which he describes the "Earth's living matter, air, oceans, and land surface as a complex system which has the capacity to keep our planet a fit place for life."

1985 - Publication of *An Environmental Agenda for the Future,* recommendations from the chief executive officers of ten major environment/conservation action groups in the United States regarding integrated, corrective actions for: protected land systems, public lands, private lands and agriculture, toxics and pollution control, nuclear issues, energy strategies, wild living resources, water resources, population growth, urban environment and international responsibilities.

1985 - Increased attention to global environmental issues during the July *United Nations World Conference on Women* in Nairobi, Kenya. Progress, during the UN Decade for Women, regarding health, sanitation, food, fuel, housing, land, human rights and population problems was evaluated within an environmental context.

1985 - Release of *A Congressional Agenda for Improved Resources and Environmental Management in the Third World: Helping Developing Countries Help Themselves* by the Environmental and Energy Study Institute, an independent, non-partisan organization which works closely with members of the Congressional Environmental and Energy Study Conference and other policy makers. Prepared by a blue-ribbon panel of corporate, development, environmental and Congressional leaders, the report proposes a new, more cost-effective approach to development aid. It stresses training and other initiatives to build local self-sufficiency, and a shift away from big, capital-intensive projects toward smaller-scale projects which efficiently use energy, water and other resources, and which are economically sustainable. [A more detailed description of the report appears later in this Chapter.]

And, in general

• Continuing attention to the African famine, with increasing understanding of the role that poor resource management, unsound agricultural and land ownership policies, unchecked population growth, and poorly planned large development schemes have played in that situation.

• Growing numbers of non-governmental organizations in all parts of the world (over 400 in Indonesia, alone) that are working locally, regionally, nationally and internationally for improved management of natural resources and prevention of pollution.

• Improved communication and cooperation between environmental and other non-governmental organizations and religious groups whose fields of interest — such as hunger, human rights, refugees, population, agriculture preservation, jobs, development and peace — are closely related.

• A parallel growth in nations with environmental agencies or ministries. In 1972, at the time of the Stockholm Conference, only about 12 countries had such agencies; now they exist in over 120 nations. The task now is to support these agencies and help assure their fuller participation in the preparation and implementation of national development plans. This may be greatly facilitated by the development of National Conservation Strategies based on the overall thrust of "sustainable development" as put forth in the World Conservation Strategy.

2. Specific Examples of Responses Triggered by Global Environmental Concerns

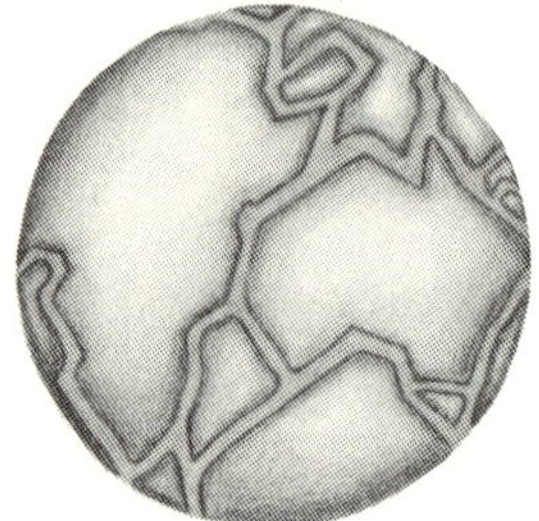

2 With that timeline of general progress as a backdrop, let's look now at some of the efforts that have come forward in response to growing concerns about the global environment. Most often concern is first raised by interested citizens, or by scientists or other specialists including members of the media. A long and tortuous delay may occur between the initial realization that a problem exists and any substantial reaction in the form of changed or new policies and programs. This delay is caused by many factors. The "new problem on the block" must vie first for visibility with the general public and national leaders. If the new problem survives this first test then it must compete for legislative or governmental attention with all the other issues that have reached this same plateau of "societal problems about which something simply must be done." Even if the issue finally reaches the point at which some legal or institutional mechanism is established in its regard, there is likely to be a continuous struggle for the resources (human, technical and financial) and time that will be needed to address the problem properly.

The first four of our examples are issues that have moved well into the public and political consciousness. Leadership has been exerted on their behalf, and strong international management mechanisms are evolving. The examples begin with a description of efforts to improve world-wide environmental data bases. This issue has been successful in gaining attention and support because there is a strong consensus on the need to provide the public and private sector with improved foundation for environmentally related decisions, and an equally strong consciousness of potential benefits to all involved. The examples then go on to explore some new mechanisms for cooperation that have been created among nations and national sub-areas in response to more difficult challenges — shared natural resources and/or transboundary pollution problems.

1. Data Base Improvement — The Earthwatch Program

Designed at the 1972 Conference on the Human Environment in Stockholm, and implemented by the United Nations Environmental Program, this international surveillance system monitors current conditions and changes in the environment, and serves as a "global early warning system." Until recently, the Earthwatch Program operated through three major components, the Global Environmental Monitoring System (GEMS), the International Register of Potentially Toxic Chemicals (IRPTC), the International Referral System (INFOTERRA), and Outer Limits, a closely related program.

GEMS works through the other U.N. agencies, enlisting their support in collection of data regarding climate, long-range transport of air pollutants, renewable resources, the oceans and human health. Programs are carried out by agencies such as UNESCO, Food and Agricultural Organization and the World Health Organization, and intergovernmental groups such as the International Union for the Conservation of Nature and Natural Resources. IRPTC uses a network of national correspondents to gather and disseminate data on the characteristics and possible environmental and health hazards posed by the thousands of chemicals produced each year. IRPTC also keeps track and shares information about the national, regional and global policies and regulatory mechanisms that are developing for the control of potentially toxic chemicals. INFOTERRA puts people with environmental problems in touch with those who may have answers through a system of national information centers in over 90 countries. Outer Limits is an international effort whose goal is to try to determine the points at which the stresses imposed on the environment by human activities will do irreparable and possibly catastrophic damage. Examples of studies under way are on the causes and effects of changes in the ozone level, and on long-range effects of weather modification through such activities as destruction of the globe's forest cover.

As experience has grown with data-gathering and monitoring of these key global conditions, both gaps and potential uses for the system have become apparent. Most of the data banks have been collected over time by specialized agencies or interest groups, and the data kept segregated by those initial categories. That meant it was not always possible for the data gathered under the different systems to be compared, or for all the data relating to a specific geographic area of the world to be looked at collectively.

During 1985 an immense step toward these important goals was taken, with the addition to the GEMS program of GRID — the Global Resources Information Database. Under GEMS, three methods of data collection are all used — imagery from satellites, observations and photos from low-level reconnaissance flights, and very detailed "ground studies." Data collected by these techniques are stored in data bases all around the world. For example, UNESCO and the World Meteorological Organization work with GEMS to maintain a World Glacier Monitoring Service located in Zurich. Data on baseline levels of key pollutants from the Background Air Pollution Monitoring Network are collected by WMO, but stored by the EPA in Washington, D.C. In Cambridge, England, the Conservation Monitoring Center, a new division of IUCN is gathering data on the status of endangered plant and animal species, on trade in endangered species and on threatened habitats. Information from these and many other data bases will be fed into GRID. GRID will locate gaps in the databanks, and will recommend that they be filled by whatever means necessary, from interpretive satellite imagery to intensive ground surveys.

Now for the near magic: every piece of information that goes into the GEMS/GRID system will be "geo-referenced" — annotated to show precisely which portion of which nation or which part of the oceans from which the data came. And because of GRID's extraordinary capacity to analyze and display information, the new system will permit the retrieval of data files or parts of them in any combination, and their analysis together or separately, to generate tables, graphs, maps or charts. This amazing capability will be of inestimable help to analysts as they examine and try to understand the problem they are investigating or the proposal they are evaluating.

The briefing document prepared by UNEP upon the announcement of the GRID system indicates how useful this type of information can be:

• Desert locusts annually devour extensive areas of carefully tended fields and pose a constant threat to food crops. These notorious insects range over 30 million square kilometers, much being land where populations depend wholly on agriculture. The area includes 55 countries, supporting 850 million people, extending east from Senegal to Bangladesh, and north from Tanzania to the southern croplands of the Soviet Union. The locust plagues originate from the relatively smaller area that nevertheless stretches for some 16 million square kilometers, from Mauritania to northwest India.

• The United Nations Food and Agriculture Organizations' Plant Protection Service has been using Advanced Very-High-Resolution Radiometer and Landsat data to locate the areas from which locust plagues erupt. Plagues occur only in ideal breeding conditions, usually after persistent rain. Females lay their eggs in sandy, moist soils and the young locusts, called "hoppers," hatch when vegetation is fresh and green. These condiions are found in only about 4 percent of the eruption belt and are virtually impossible to locate effectively with conventional ground and aerial methods.

• Landsat and Advanced Very-High-Resolution Radiometer data have been used to pinpoint areas which are potential breeding sites, by locating the vegetation blooms. Once the sites are located, ground checks determine if there are hoppers present. If there are, they are eradicated by spraying. The environment benefits in two ways: the locusts are removed, and much less insecticide is used than with traditional blanket spraying. This reduces costs and pollution.

In the same briefing document a summary of potential benefits to the world stated that "on a world-wide level GRID will be used to make statements about global trends, such as changes in climate and increasing desertification. For individual nations, GRID will be valuable at any scale of planning, from national agricultural schemes to the siting of roads and human settlements."

The briefing document went on the say, "Perhaps GRID is best summed up in terms of a medical analogy. When people see their familiar world in disarray they are often forced to stand by, unable to understand what is wrong, forced to watch helplessly. Like a vet faced with a sick animal, environmental managers cannot ask forests and seas what is wrong when they show signs of ill health. With GRID, the environment will begin to resemble a human patient, capable of answering questions about itself, helping doctors to locate the problems and to treat them. When that happens, we will be able to move from crisis management to prevention, from a sick environment to a healthy one."

The United States has played a key role in the initiation of the GRID program. The equipment for GRID has been provided through donations from a number of national agencies and private companies, including the National Aeronautical and Space Administration and the Prime Computer Corporation which, between them, have provided most of the hardware. The software that has been donated includes the NASA's Earth Resources Laboratory Applications Software, and ARC/INFO from the Environmental Systems Research Institute in Redlands, California.

2. UNEP's Regional Seas Program — Achieve Common Environmental Goals.

The Regional Seas Program of the United Nations Environment Program steps into more difficult challenges than those involved in improving the global data bases: it focuses on environmental problems in coastal and marine areas that come under the jurisdiction of adjoin ing nations. Carried out in collaboration with 14 United Nations agencies, 12 other international organizations and participating national governments, it is now stimulating cooperation among and helping create improved environmental management within more than 120 nations around the globe. The program is a direct outgrowth of the 1972 U.N. Conference on the Human Environment, in Stockholm, when declining fisheries, contaminated drinking water and filthy coastlines in numerous parts of the world, motivated world leaders to agree that coastal areas were important targets for early international environmental cooperation.

What causes such severe stress on coastal areas?

Most of the world's most populous cities are located in coastal areas or estuaries (where fresh water from rivers gradually mixes with salt water of the oceans). Major cities have developed in these sites because of the availability of water supplies, marine food resources and the potential for water-borne commerce. Industry, shipping, tourism, fishing, human settlements and agriculture have developed intermixed, creating a high concentration of economic activity, available services and employment opportunities. The result? Seven out of every 10 people on Earth live within 50 miles of one coastal area or another.

These concentrations of people and activities place considerable stress on the rich and fragile natural systems of the nearby sea-zone areas where 20,000 of the known varieties of fish, 30,000 species of mollusc and almost all crustaceans live — species which collectively provide all but 10 percent of the world's seafood catch.

Coastal areas are also the natural habitat for a wide variety of plant, animal and bird life. Some of these species are harvested directly to provide food for humans or animals. Even those that escape this fate may be affected if the human settlements and activities are not well-managed. Oil spills, pollution from land-based activities (agricultural runoff, industrial wastes and municipal sewage — which account for 80 percent of all the pollution that reaches coastal waters), construction, mining, dredging, dredge spoil disposal and other activities can all contribute to the destruction of habitats, or even to periodic fish or wildfowl kills.

The worst first

The Mediterranean Sea was chosen, in 1974, as the first area in which to apply UNEP's Regional Seas concept. Mutual concern over degradation of that water body was enough to bring together nations as culturally and politically different as Algeria, Cyprus, Egypt, France, Greece, Israel, Italy, Lebanon, Libyan Arab Jamshiriya, Malta, Monaco, Morocco, Spain, Syrian Arab Republic, Tunisia, Turkey and Yugoslavia (plus the European Economic Commission) to work out ways to reduce pollution and improve protection and management of shared coastal and marine resources.

Now, 13 years later, Albania is the only Mediterranean nation that is not a participating party. All others have adopted the comprehensive Action Plan, and formal legal agreements that deal with: dumping, cooperation in case of emergencies (like oil spills), land based sources of pollution and protected areas. Over 80 national scientific institutions, backed up by FAO, WHO, UNESCO and other agencies of the U.N. system, participate in cooperative field research and monitoring; a 24-hour contingency oil-spill center is operational in Malta; and national development plans are a basis for regional consultation under the plan. UNEP continues to provide a coordinating secretariat in the service of the contracting parties, but the entire enterprise is governed and financed by the parties themselves.

Regional Seas Programs have been developed in nine other coastal areas around the world, and consultations are under way to see whether an 11th Regional Sea Program for South Asia (which would cover the coastal areas of Bangladesh, India, Maldives, Pakistan and Sri Lanka) will come into being.

The evolution of an operational regional seas program

A program may take years to become fully operational. UNEP has no regulatory authority and cannot initiate action toward any such program on its own. The process can begin only when UNEP is asked for assistance. UNEP, working with selected U.N. and other organizations, provides seed money, or catalytic financing, in the early stages. Working with scientists from all interested nations and a host of international governmental and non-governmental organizations, overviews are prepared on the nature and extent of the problems and opportunities a regional program might address. Considerable effort goes into determining the particular interests of each party, and the scope and substance each would like to see in a plan suitable for its coastal region.

The program enters an operational stage only when an action plan is formally adopted by an intergovernmental meeting of the governments in that region. Each action plan, although different in detail, includes the following components:

• Environmental assessment - descriptions of how the causes of environmental problems and their magnitude and impacts will be evaluated;

• Environmental management - identification of types of cooperative regional activities through which the identified problems might be addressed;

• Environmental legislation - an umbrella regional convention elaborated by specific technical protocols to provide the legal framework for regional and national action;

• Institutional arrangements - designation of an interim or permanent secretariat for the action plan; agreements on frequency of intergovernmental meetings to review progress, approve activities and provide budgetary support;

• Financial arrangements - through which the governments of the region progressively assume full financial responsibility for the Regional Seas Program (usually through the establishment of a regional trust fund to which governments contribute, and which is managed by the designated secretariat).

Approval of the action plan comes first, later followed by the legally binding convention and specific technical protocols.

The wider Caribbean :The regional sea program closest to North America

The effort to develop a Regional Seas Program for the Wider Caribbean began in 1977. This complex area is bound by one of the world's most highly industrialized coastlines (the southern Gulf of the United States) and by several major producers of offshore oil, including Mexico and Venezuela. It also includes two of the planet's most valuable and threatened natural systems: coral reefs, which support an immensely diverse range of species; and mangrove swamps, which serve as nursery areas for many marine species.

Development pressures in the area, combined with the extreme fragility of the ecosystems involved, made it vital that an action plan be based on sound environment-development relationships. As background for deliberations on an action plan, overviews were prepared for nine selected issues upon which the future development potential of the region depended. The issues examined were: freshwater resources (for agriculture as well as drinking) and health-related diseases; agricultural resources (soils, pasturelands, forests and agricultural chemicals); marine and fisheries resources; wildlife and genetic resources; energy resources (including oil, gas, biomass, coal, hydropower, geothermal and solar); mineral resources (including oil); human settlements; tourism; and transportation.

These reports, combined with other types of baseline information, led to the adoption, in 1981, of an Action Plan for the Wider Caribbean that addresses environmental protection and comprehensive management of the area's resources in a way designed to meet economic development and environmental objectives. The action plan addresses such diverse but interrelated issues as: alternative sources of energy (to reduce the rate of deforestation); oil spills and marine pollution; protection of coral reefs, mangroves, coastal lagoons and turtle-grass beds; fisheries development, including fish farming; an early-warning system for such natural disasters as hurricanes; drinking water; environmental health problems; housing; tropical forest and mineral resource management; soil use and erosion control; agriculture; and, the equipping of science laboratories in the less developed nations.

In total, 66 environmental projects are included in the action plan. Programs have been initiated on eight, including oil-spill control, watershed management, public awareness and environmental impact assessments.

In 1983 a framework for legal action, the Cartagena Convention, and a protocol on Cooperation in Combating Oil Spills were put forward. Seventeen of the 23 states involved, including Columbia, Cuba, France, Grenada, Mexico, Nicaragua, the United Kingdom, the United States and Venezuela have adopted the measures — countries who, in some cases, find little else to agree upon. The convention calls for the "Protection and Development of the Marine Environment of the Wider Caribbean Region" and binds ratifying states to "prevent, reduce and combat pollution and ensure sound environmental management."

A major reason for the success of the regional seas program: Its flexibility

One of the major factors behind the success of UNEP's Regional Seas Program is the way each has been allowed to develop in ways that reflect the precise needs of the area. In the Mediterranean, the first priority of the participating nations was to try to slow the rate of degradation. In the Wider Caribbean, the Regional Seas Program is serving as a basis for integrated planning that addresses both development and environment needs. This approach may help eliminate some of the root causes of the many problems that are presently manifesting themselves in ever-increasing numbers of refugees, widespread political instability and vulnerability to subversion.

But the Regional Seas Program for the South Pacific is planned in such a way as to aid nations in addressing a very different challenge — less to change traditional behaviors and attitudes than to maintain them. As a report, prepared for the 1982 Raratonga meeting on the "State of the Environment in the South Pacific," noted:

"Environmental management is not a new concept for Pacific peoples. Wherever natural resource management was needed the traditional cultures of the region developed practices which protected their essential interests. These included land and reef tenure systems; permanent and temporary taboos on species or places; refined and selected fishing techniques; agroforestry, terracing and irrigation; windbreaks, bush fallow, and other agricultural and soil management practices, etc. The cultural heritage of the Pacific is full of examples of sound environmental management that is equivalent or even superior to modern methods. One of the great tragedies of this region is that this heritage is rapidly being lost just as the need for it is increasingly apparent."

3. Coping with Transboundary Environmental Issues on the New England-Canadian Border

Acid rain and water pollution are the U.S.-Canadian environmental issues that have received most publicity and governmental attention in recent years. A detailed look at one section of the U.S.-Canadian border — the New England Region — shows the types of new mechanisms for collaboration that are resulting in areas where the environmental consciousness of political leaders on both sides of the border is very well developed.

As the decade of the 1980s began, New England, like most parts of the United States, had fairly sizeable institutions whose function was to bring the states together to address shared economic and environmental problems and goals. Funded jointly by the federal government and the states, NERCOM (the New England Regional Commission) was the primary mechanism for dealing with economic issues, and NERBC (the New England River Basins Commission) served as the vehicle for cooperation on regional land- and water-related issues. In early federal cost-cutting measures both NERCOM and NERBC were terminated in the fall of 1981.

Using leftover funds from the two organizations, plus state contributions, an informal body which had been established in 1936, prior to the existence of either NERCOM or NERBC — the New England Governors'

Conference — was formalized into a non-profit organization and charged with the task of maintaining communication and coordination among the New England governors and with the premiers of eastern Canada on issues of common concern. At the urging of the 20 citizens who served on the NERBC Steering Committee on Public Involvement, some of the leftover NERBC funds were used to maintain a Water Resources Program within the New England Governors' Conference. The water program was directed to maintain the useful communication and coordination among the multiple agencies within each state and the departments within federal agencies with water-related mandates. (On the federal side alone that list includes the Departments of Agriculture; Interior — which, in the eastern sector of the country operates through the U.S. Geological Survey, Bureau of Land Management, Fish and Wildlife Service, National Park Service — Energy; Housing and Urban Development; Commerce; Transportation — along with the U.S. Army Corps of Engineers, the Environmental Protection Agency, Federal Emergency Management Agency and the Federal Energy Regulatory Commission.)

Over the past five years a quite exciting set of initiatives has evolved. For the first three years of the Governors' Conference's existence, its Federal/State Water Resources Steering Committee, composed of the chief state planning officials and representatives of the regional offices of the cited federal agencies, prepared updated "Priorities for Federal Assistance." Funding and/or technical assistance with the region's most serious water-related problems was requested from Congress, the administration, and the appropriate federal agencies. Each year these recommendations were reviewed and then approved, sometimes with modification, by all six New England governors acting in unison through the Conference. In 1985, recognizing the increasingly strong role of the states, the Conference, and others in improved water management, a revised report was prepared, the first annual Action Plan for Water Management in the NE/NY Region. The plan outlines recommendations for action by the governors, state legislators, the New England

Governors' Conference and the Conference of the New England Governors and Eastern Canadian Premiers, as well as by the administration, congress, and federal agencies. The 1986 action plan has four major focuses:

• Improved regional management of toxic substances (including low-level radioactive wastes and municipal solid wastes);

• Reduced regional damage from acid rain;

• Improved management of regional groundwater resources;

• Maintained excellence in other key areas of water management (including water supply, waste treatment, flood protection, navigation, agriculture, water-oriented tourism and recreation, energy development, drought preparedness)

To the Water Resource Steering Committtee, the New England Governors' Conference original environment-related group, have been added several new committees to work on the increasing numbers of environmental problems that cross state and/or the U.S.-Canadian borders. The new bodies include:

• An NEGC Environment Committee, which is composed of the top environmental appointee in each of the New England states. The group meets regularly to exchange information on shared problems and to advise the governors. One of the major focuses of this committee at the present time is evaluation of present state programs for solid waste management, and determining whether further federal involvement is needed.

• An Environment Committee of the Conference of New England Governors and Eastern Canadian Premiers, which is composed of the NEGC Environment Committee members and their counterparts in the eastern Canadian provinces. This group helped develop an Acid Rain Reduction Plan, since adopted by the governors and premiers, which is designed to reduce sulphur dioxide emissions in the New England-eastern Canada region 32 percent by the year 1995. Recently, Canada has launched on its own a major — and expensive — program to cut emissions by its own polluters. With these actions having been taken it will come as no surprise that interest among Canadian and New England elected officials is very high regarding the recently released *"Joint Report of the Special Envoys on Acid Rain,"* by former U.S. Transportation Secretary Drew Lewis and former Ontario Premier William Davis. The report calls for the United States to launch a five-year, $5 billion program for the commercial development of new technologies to cut emissions from coal-fired power plants (which are understood to be the main cause of acid rain in the United States) and of related problems in Canada. While the call for new technologies — as opposed to the continuation of research alone — is welcomed, the report is the first of three such studies within the past five years that has not recommended use of available technologies to cut emissions. It has been estimated that acid rain emissions could be cut 50 percent by the application of such technologies. The estimated average cost, which would be added on to household utility bills, is anticipated to be 60 cents to $4.50 a month, the precise amount depending on location. The response of the U.S. to the report is not yet clear; nor is the issue of how such a program to create new technologies, if approved, will be funded.

• A Forest Productivity Working Group of the Conference of New England Governors and Eastern Canadian Premiers. Participating on this body are the chief forestry managers of each New England state and eastern Canadian province. The group meets regularly to discuss common problems and has input into the development of policies, like the Acid Rain Reduction Plan, which will help to protect the valuable forest resources which cover approximately 447,000 square miles, or 57 percent of the region's land. In this two-nation area, over 214,000 people are employed directly by the forest products industry and an additional 183,000 jobs depend on it indirectly. The forest products industry contributes nearly $8 billion annually to the regional economy. When the jobs and economic activity stemming from tourism and recreation in the region's forests are included, the forest resources is clearly seen as the basis for a major part of the productivity of New England and eastern Canada.

Other shared regional problems that the New England Governors' Conference and the Conference of New England Governors and Eastern Canadian Premiers are looking at relate to tourism; fisheries resources; hydropower development; electrical energy transmission and sale; ports management; protection of small-scale agriculture; and other environmentally related issues like the U.S. national search for suitable disposal sites for high-level radioactive waste materials.

Within the past five years the attention given to regional environmental concerns by the governors and the premiers has increased steadily. Recognizing the strong link between protection and wise management of the region's resources and the strength and vitality of its economy, stronger and stronger frameworks for decisions are being developed within each state, among the states through the New England Governors' Conference, and with the eastern Canadian provinces. Still another complex and particularly intriguing issue, however, is waiting in the wings — the Canadian proposal to develop a major tidal power facility in the Bay of Fundy.

It will come as little surprise to anyone that the New England region experiences winters. In contrast to some parts of the United States, the period of cold weather, often lasting from October until April, means that more than 30 percent of a family's income may be absorbed by household fuel costs. High heating costs are likewise a factor for every public building. Heating and energy production costs affect the extensive industrial and commercial development in this region. The energy costs of living, providing public services and doing business in New England are considerable — and the search for ways to hold these costs down, in order to make the region's enterprises competitive on a national and international scale, never ends.

Moreover, to the extent that this energy is produced from fuels imported from other nations, the region is vulnerable to the political situation in the oil-producing nations, and affected by changes in other costs, like shipping. The term "energy independence" has real meaning to this sector of the nation. At the same time, there is also genuine concern within the region about the effects of the region's present energy mix; acid rain from oil- and coal-powered plants both outside and within the region ; thermal pollution, public safety and radioactive waste disposal questions associated with nuclear power plants; affects of major hydropower facilities on fisheries resources and water-oriented recreation; and the erosion, loss of wildlife habitat and poorer air quality in some sectors that is resulting from expanding use of wood as a heating fuel.

Thus, when a neighboring nation with whom political relations are good proposes to develop a facility on their "turf" that will produce a stable and less expensive source of electricity which New England could import without adding to the air or water pollution, natural resource or public safety concerns associated with other energy sources, that should be wonderful news. Right?

The Canadian proposal, if constructed as presently designed, would produce electricity by building a five-mile-long dam across Midas Bay in the Bay of Fundy that would capture part of the tremendous power created by one of the world's greatest tidal differentials. At the head of the Bay of Fundy this rise and fall is in the range of 45-50 feet. Tidal power has much to recommend it. It is inexpensive; totally renewable; "clean" in the sense that it releases no air or water pollutants, or waste heat, and no radioactive wastes. Obviously, there are localized impacts during and from construction of such major facilities, and on the land area where the water would be temporarily impounded before being released to generate electrical energy. But on the surface of things the Fundy Bay project would appear to be an ideal way for New England to help decrease its dependence on imports from less stable parts of the world, and to hold electricity costs down by using a cheaper source and, at the same time, be able to avoid increasing the environmental and public safety aspects of the region's existing energy mix. The proposal would also be of economic benefit to Canada, who would look to sell to the northeastern states upwards of 90 percent of the electricity generated.

All of the above may turn out to be absolutely true, but a few additional issues need to be clarified before the balance sheet can be totaled. An early numerical model that examined and described the tidal properties of the Gulf of Maine and Bay of Fundy indicated that the modification of the tidal pattern created by the proposed tidal power project could affect tides as far south as Provincetown, Massachusetts, on Cape Cod. The study also indicated that changes will be created in the existing circulation patterns of the ocean as it flows near the facility, which is the route taken by species of migratory fish that sweep up from more southerly waters, traveling all the way to Fundy Bay before turning toward the land mass and coming down along the Canadian-New England coast to their respective spawning and nursery areas.

The dimensions of these two effects must be added to the equation that evaluates the trade-offs surrounding the Fundy Bay project:

• Will there be impacts on fisheries resources? (A question of great importance to both Canadian and U.S. interests.) And if harmful, can these be modified through changes in design and operation of the facility?

• What size tidal change can be expected at points along the Canadian-to-Maine-to-Massachusetts coast?

Why does this matter? If the differential is small — literally only a few inches of change in the present high and low tides, then negative impacts may well be minimal. If, however, the differential in the present high and low tides is sizeable, the impacts could be: changed dimensions of wetlands; increased vulnerability to coastal and island flooding; increased vulnerability to coastal erosion changes in the depths of existing harbors and navigation channels, which would affect boating and shipping safety and the size ships which individual harbors could accommodate. All of the above could impact coastal industry and commercial investments, private investments as well as tourism and water-oriented recreation, which are likewise mainstays of the New England economy, changes in habitat for wildlife, birds, shellfish and non-migratory fish species.

Fortunately, the scientific and technical capability to assess many of these questions exists, and funds are being sought from the U.S. Congress to simulate the tidal patterns and provide some answers. If the impacts are predicted to be minimal, then the issue will depend on success with finding financial backers and arranging suitable legal arrangements regarding the sale of the electricity generated. If the impacts are predicted to be major, the situation will take on a dramatically new twist. For there is no existing bilateral or international legal agreement that would require Canada to do anything (such as changing the design or size of the facility) to minimize the impacts on U.S. natural resources and existing investments in homes, roads, flood control or navigation facilities, industry, businesses, etc. Interestingly enough there is an existing U.S.-Canadian agreement under which the United States has some control over overland transmission lines that are used to bring Canadian electrical power into this country, but none that would protect this wider range of U.S. marine and coastal interests. As the situation stands now the United States, and particularly the New England coastal states, would have to rely upon "Canadian good will."

4. Multi-National Cooperation in the Protection of Endangered Species

This example in the subcategory of environmental issues for which good management mechanisms are evolving involves efforts to regulate the huge global trade in animals and wildlife products such as furs and ivory. In 1973 an international meeting convened in Washington, D.C., and created CITES: the Convention on the International Trade in Endangered Species of Fauna and Flora. The aim was to prevent endangered and threatened species from becoming extinct through commercial overexploitation. Over 80 nations have now joined in this effort. CITES meets every two years to review wildlife trade regulations and the overall implementation of the convention.

CITES is based on lists that segregate species according to levels of threat. For example, placement of a species on Appendix I allows the proposing countries to establish a strictly set and managed quota of the numbers of that species that can be hunted or traded. The quota is reviewed at the next two-year meeting to see if it is achieving the mutual goals of all parties concerned. This provides a mechanism through which nations that have different perspectives on aspects of a trade problem can come to agreement. For example, at the 1983 CITES meeting, the United Kingdom, in an effort to ease the task of monitoring the ivory trade, proposed making unnecessary the import licensing of items made of ivory that weighed less than .5 kilograms. During the discussion this proposal was withdrawn when India showed that such a move could further endanger populations of the Asian elephant, which is already under severe threat.

Like all management mechanisms that are effective over time, CITES monitors its own progress and adjusts its emphasis as necessary. As destruction of vegetation throughout the world accelerated, the need to use CITES and other mechanisms for protecting genetic diversity became even clearer than before. When it became apparent that, of the 25,000 plants threatened, not all were subject to trade (many are threatened by collectors) several dozen plant species, mostly cacti, were added into protected status under CITES. In addition, the World Wildlife Fund/ International and the International Union for the Conservation of Nature and Natural Resources set up a plant campaign for 1984 aimed at showing the crucial importance of plants to man, and to the Earth's life support mechanisms.

Sometimes what is needed to protect a species of plant or animal turns out to be, not convincing people of the importance of the species to their lives, but just the reverse. The black rhino is a case in point. In Asia, rhino horn has long been prized as a pharmaceutical, particularly for reducing fever. Concerned about the plight of the black rhino, whose numbers have plummeted to some 10-15,000, WWF and IUCN commissioned a study by the international pharmaceutical firm, Hoffman-LaRoche & Co. The study found that, like human fingernails, the rhino horn is made of agglutinated hair and that there was no evidence that it has any medicinal effect. Dr. Arne Schiotz, WWF Director of Conservation, noted, "This proves that rhino horn is of no use to anyone except the original owner. One would get the same (medicinal) effect from chewing one's own fingernails."

Steps are also being made to deal with the other major threat to the black rhino which stems from a quite different tradition. In the last decade over 90 percent of the most concentrated populations of black rhino, in Kenya, Uganda and northern Tanzania, have been slaughtered. The Yemen Arab Republic was formerly the world's single biggest market for poacher-supplied horns, which were carved into elaborate dagger-handles. Considered a status symbol, these *janbiyyas* are proudly worn at the waist by 80 percent of the adult Yemen males, and may cost up to a thousand dollars. In 1983 the Yemen Arab Republic banned the import of the animal's horn, noting that the government's decision was "in keeping with its endeavors to protect endangered species throughout the world and that the government of Yemen fully supports the conservation activities of WWF and IUCN which have drawn attention to such important issues."

5. Protecting the Antarctic Region Resources

This selection falls midway between our two perspectives of serious environmental issues for which strong management mechanisms have or have not evolved. It is excerpted from a guest editorial by Dr. Gerard A. Bertrand, who is the president of the Massachusetts Audubon Society, Lincoln, Massachusetts, and former director of the Office of International Affairs, U.S. Fish and Wildlife Service, Department of the Interior. Dr. Bertrand discusses this important region of the world for which a protective mechanism is in place — but one which runs out in a few years and for which renegotiation must be attempted. In the interim since the protective mechanism was established, however, a lot — including the Falklands Islands War — has taken place. A peaceful long-term solution regarding this example may require the "marrying" of two international agreements.

(Published in 1983 in Volume 3, Issue 1 of The Environmentalist, *by Dr. Gerard A. Bertrand.)*

Three South American countries — Chile, Argentina and Brazil— plus several from northern Europe — the United Kingdom, Norway, France and others — hold claims to sections of the Antarctic. Even before much was known about the area, this jumble of claims held such potential for serious conflict that in 1959 the United States led the way toward the establishment of the Antarctic Treaty, an international agreement that involves both "claiming nations" and others, and which has held the claims of individual countries in abeyance for more than 20 years. But the treaty expires in 1991.

In the years since the treaty was adopted, the realization that the Antarctic contains a wealth of living resources (including krill, fish, whales and seals) as well as the potential for enormous mineral resources has caused countries to harden their stances on national claims. One can easily see the potential for conflict — the territory claimed by the United Kingdom entirely envelops Argentina's claim, as does the British claim to the subcontinental islands, including the Falklands — the scene of that rather bizarre mini-war not long ago. In addition, the decision of the United States not to sign the Law of the Sea Treaty because of the seabed mining issue has raised fears among Antarctic Treaty organizations that the United States' intentions for the Antarctic may be far from benign.

Few people realize that the Falkland Islands, including the Sandwich Islands and South Georgia (all subject to claims by both Britain and Argentina) contain one of the greatest wildlife concentrations on Earth. Many have referred to the Falkland archipelago as the Galapagos of the Antarctic. The islands cover over 3 million acres of land (approximately 4,600 square miles — less than half the size of Belgium). There are 63 bird species including five species of penguins. Penguins and sheep share the same fields — a unique situation which exists nowhere else on Earth. Albatross, shearwaters and the endemic Falkland Island Flightless Steamer Duck are all found in abundance. Besides the spectacular birds, elephant seals over 20 feet in length can be approached on the beaches while minke and fin whales as well as killer whales abound in the waters surrounding South Georgia and the South Sandwich Islands. The marine life generally is among the richest on the planet.

In short, the Falkland Islands with their uplifted terraces and cliffs are beautiful in their wildness and are an ideal candidate for protection because of their ecological richness. Militarily and politically, however, their situation is dangerous and complicated. The Falkland Islands, traditionally staging areas and ice-free ports for exploitation of the whales, seals and birds of the Antarctic region, could also perform such functions for offshore development or mineral exploitation. This locational issue was a very important aspect of the seriousness of the U.K.-Argentine conflict. However, the islands have a far greater long-term value as a base for sustained production of renewable resources than short-term destructive exploitation by present generations. If there are minerals, gas and oil in the region they could be held as reserves for mankind's long-term future needs. In the meantime, preservation of the living resources and ecosystems of the Falklands would serve present global political, economic and ecological needs.

The World Heritage Convention, still another existing international agreement, may provide a way to avoid further conflict in the area. Convened in Paris in 1972, at the instigation of the United States, the Convention now has more than 65 signatories including Argentina. The World Heritage Convention, managed by the Director-General of the United Nations Education and Scientific Organization, protects areas of outstanding natural and cultural value throughout the world. For example, it protects the pyramids of Egypt, the palace at Versailles, Olduvai Gorge in Kenya, the Galapagos Islands, and the Great Barrier Reef. Declaring the Falklands an international park (environmental zone or territory) under the World Heritage Convention administered under a U.N. mandate would alleviate the British burden of enormous annual military maintenance costs involved in guarding those islands and automatically would remove the irritation of the British presence for the Argentines. Falkland Islanders could maintain their choice of citizenship and lifetime residency — just as is done for inholdings in many national parks. Such a solution could permit the continuation of many of the present uses of the islands by residents, such as sheep farming and fishing. These uses have continued for more than 200 years, and, along with the people themselves, have accommodated to the wildlife on the islands.

Declaring a Falklands International Environmental Territory under U.N. administration would go a very long way toward achieving the international harmony necessary for renegotiating the Antarctic Treaty Convention in 1991. It could lay a basis for avoiding a conflict which could again pit the Antarctic Treaty nations, including the Soviet Union and the United States, against each other in the grab for potential Antarctic resources. In such a situation the World Heritage Convention may be able to provide the sanity called for by this volatile situation. Learning from the high costs of the "restricted war" that recently took place in this South Atlantic region, the establishment of a Falklands International Environmental Territory under the World Heritage Convention and a renegotiated Antarctic Treaty Convention could assure that the area remains peaceful for the foreseeable future.

6. Improved Management of Toxic Substances

There are many reasons why the United States is presently faced with multiple and very difficult challenges relating to the use and disposal of toxic substances. The nation's advanced medical technology and medical research and development activities produce chemical and low-level nuclear wastes. Our highly developed industrial sector, and widely available services — like dry cleaning and photographic film development — produce a plethora of wastes that can affect human health and pollute air, surface water, groundwater, and soils. Our modern agricultural technology, with its dependence on herbicides and pesticides, can leave pollutant trails in air, water and soils, as well as in animal and human food chains. Sewage sludge can be a source of unwanted toxics — toxics made more concentrated by the very processes that treat the sewage. Nuclear power plants and research leave us with low-level radioactive wastes to worry about; and military research, especially in the nuclear weapons field, likewise adds to the total of what must become safely used and disposed of. Even household wastes — cleaning fluids, old paint and hundreds of other products used in our daily lives — contain substances that, improperly disposed of, can create human health hazards or threats to key natural systems like groundwater and wetlands.

One of the toughest aspects of this issue is that many of the decisions about revised management processes must be taken at the local governmental level, where elected and appointed officials and citizen volunteers must try to balance the needs associated with protecting public health, safety, property values and the local natural resources with the equally pressing (and often more vocal) needs to maintain jobs and protect the local economy. The search for safer management measures cannot even stop with our own community. Toxic substances and/or waste products that are generated elsewhere but transported through our communities create another whole set of concerns, as does the worry of whether our community is "down-wind" of a potential source of toxic air pollution.

The problems reach out across our national borders. For instance, in efforts to prevent the release of hazardous materials along the 2,000 mile inland border between the United States and Mexico, a Joint Inland Contingency Plan providing for a mutual response program for hazardous material spills, has been approved — with a pilot program for the Calexico-Mexicali border area initiated in 1985. Moreover, the two countries have agreed to address "the potential for indiscriminate and uncontrolled transborder area movement and disposal of hazardous materials in the border area" by developing a "hot line" communications link — a process for rapid exchange of information on transborder movement of hazardous wastes, a coordinated regulatory program, and a pilot, bilateral training program in 1985/1986 for U.S. and Mexican customs officials.

But the interests of U.S. citizens cannot be met by improved toxics management measures in this country and along its borders. A recent U.S. Public Health Service study found that 40 percent of all the coffee beans imported into the United States were contaminated with pesticides whose use is banned in this country. Thus worldwide cooperation on safe use and disposal of toxics is the only way that human health within any nation, and the well-being of vulnerable parts of global natural systems — bees which are the sole source of pollenation for many crops; birds and insects which are natural pest controls; and soils and water without which crops cannot be produced — can likewise be protected.

Within the United States the search for safer methods for use and disposal of all of these substances has been under way since early in the 1970s, and some key pieces of national legislation, like Superfund, have come into being. As the importance of improved management of these substances has become clearer, some very exciting innovation within industry has begun the shift toward not only safer storage, use and disposal of toxics, but redesigned manufacturing processes that are reducing the waste stream through recycling and/or substitution of less-hazardous chemicals. These complex issues which push at society with such urgency have created an immense need for communication and collaboraion among policy makers, lawyers, industrial experts, researchers, elected officials, citizen leaders, members of the media, educators and the public. In response to this aspect of the problem of toxic substance management, some new institutions have been brought into being to help meet this vital need. Descriptions of two such institutions follow:

• The Public Policy and Education Division of the University-Industry Cooperative Center for Research in Toxic and Hazardous Materials, Rutgers University, New Brunswick, New Jersey. The center is a consortium of Rutgers, the New Jersey Institute of Technology, the University of Medicine and Dentistry of New Jersey, and Princeton University. The cooperative center is funded by the National Science Foundation, industry, and the state of New Jersey. The 1984-85 budget included over $1.5 million which funded 21 research projects in the areas of: incineration, biological and chemical treatment; physical treatment; site assessment and remediation; health effects; and public policy and education.

Two of the funded projects were in the Public Policy and Education Division. One is analyzing public participation in siting hazardous-waste management facilities. The project includes a national survey of public participation, case studies of processes that offer the best hope of resolving siting issues, and a national conference on public participation.

The second project provides continuing education about risk assessment for journalists. The research project is studying the elements of good and bad reporting of environmental health stories and the feasibility of providing a team that would aid journalists during public health emergencies. Lastly, education programs for journalists have been introduced. For example, a hazardous-materials spill incident was simulated, videotaped and turned into an educational tool for journalists.

• The Center for Environmental Management, Tufts University, Medford, Massachusetts, has received a $1 million federal research grant to conduct seven projects related to the critical problems caused by solid and hazardous wastes. The goal of the center is to protect human health and the quality of life through development of:

— A better understanding of the health and environmental effects of chemical, biological and physical agents in the environment;

— Technologies, strategies and policies to minimize or eliminate the production of wastes and other environmental toxins, and to manage the remaining products in a safe manner; and

— Improved programs for interaction among, communication between and education of government, industry, faculty, students, the public and the news media.

Current CEM projects include research into: source reduction; extending the useful life of sanitary landfills;

hazardous waste from educational institutions; state experiences in dump site cleanup; plus a number of technical research studies. The center has also received special funding for an Asbestos Information Center. Tufts is one of only three institutions nationwide to be selected by EPA to develop and implement pilot education programs and information services aimed at school and health officers, abatement contractors, and building owners and managers who are in a position to limit the public's exposure to hazardous asbestos.

A number of companies are working with the CEM to assist in the development of a partnership with industry. The companies include IBM; Monsanto; Dow Chemical; Camp Dresser & McKee; AT&T; Rhom & Haas; Polaroid; General Electric; New England Mutual Life; Associated Industries of Massachusetts; Digital Equipment; The Foxboro Co.; Boston Edison; Environmental Research and Technology; and Dennison Manufacturing Corp. Stone & Webster Corp. of Boston, one of the world's largest consulting, engineering and construction firms, is helping through a "loaned executive" program. An environmental engineer with 12 years of experience in impact assessment and construction management will be working with CEM staff to coordinate the center's long-range programs with the needs of industry, and to further develop the industry/university partnership to guide and support the center's programs.

7. Moving Toward Ozone Layer Protection

Our next example of an environmental problem for which management mechanisms are just beginning to evolve is one that has been under scrutiny for the past 15 years. The ozone layer is a belt of rarefied gas, mainly found 10-50 kilometers above the Earth's surface. The layer shields the planet from the sun's deadly ultraviolet-B radiation. In the early 1970s scientists feared that the Earth's blanket of ozone was being depleted, and that severe health hazards to humans and other living things could result, including increases in skin cancer, reduced agricultural productivity and the possibility of impaired fish yields. Based on these concerns, the 58 nation members of the governing council of the United Nations Environment Program decided to initiate work toward establishing a plan to protect the ozone layer.

Although much still needs to be learned, by now experts generally accept that for every 1 percent depletion of the ozone layer, greater penetration of ultraviolet B radiation could increase the incidence of non-melanomac skin cancer in humans by 4 percent. In wheat, rice, soya beans, barley, potatoes and beans, growth productivity and other functions could be impaired at that same level of decrease. A larger ozone decrease, of 5-25 percent, could damage the larvae of fish, shrimps and crabs, and the zoo-plankton and phyto-plankton so essential to the aquatic food chain.

By early 1985 the preliminary work had been completed toward a convention that would mark the beginning of cooperative international efforts to protect the ozone layer. And in March of that year, a conference in Vienna, Austria, drew together international leaders to finalize and sign the convention.

In addressing the opening session of the conference Dr. Mostafa K. Tolba, UNEP Executive Director, spoke of the uniqueness of the Global Convention on the Protection of the Ozone Layer. He noted that "there is nothing 'regional' about the convention (as with those established to protect specific geographic areas like regional seas), for the ozone layer protects every square meter of our planet and therefore every person from every continent and country." Dr. Tolba went on to say, "the atmosphere that we hope to save from potentially irreversible imbalance makes up an entire component of our environment — it is as if we hoped, with one agreement, to prevent permanent damage to all the oceans or all the soils in the world." Dr. Tolba also called attention to the fact that this was the first international convention to address an issue that "seems far in the future and is of unknown proportions." He described the convention as the "essence of the 'anticipatory' response so many environmental issues call for: to deal with the threat of the problem before we have to deal with the problem itself." The effort marked, in Dr. Tolba's opinion, "a sign of the political maturity that has developed over the years, which recognizes how vital it is that we act to prevent environmental degradation or disaster with wisdom and foresight."

The convention was adopted by the 41 countries represented, and by the European Economic Commission; but the specific protocol to the convention to control chlorofluorocarbons was not adopted. Chlorofluorocarbons — particularly CFC 11 and CFC 12 which are used extensively as aerosol propellants, refrigerants, in synthetic foams and industrial applications — are believed to have great potential for destroying ozone.

The conference directed UNEP to continue to prepare the protocol for eventual adoption, and called on national governments to do all that they can individually to control chlorofluorocarbon emissions in the meantime.

Human-induced ozone pollution: another issue

Preservation of the natural ozone layer in the upper atmosphere that protects Earth from solar radiation has to be distinguished from the problem of human-induced ozone pollution that is also affecting some sections of the United States.

This polluting, as opposed to protecting, form of ozone can be generated when the burning of fossil fuels releases nitrogen oxide gases into the air. Depending upon which chemical transformations take place, acid rain or ozone can result. A chemical reaction, involving sunlight and nitrogen oxides, produces ozone — a major component of smog. In the eastern United States it has been determined that 40 percent of the nitrogen oxides released into the air come from automobile exhausts, 30 percent from power plant emissions and the remainder from home furnaces and industrial boilers. Tests to attempt to determine impacts of human-induced ozone have been funded by EPA. Thus far, tests have been carried out on fields of soybeans, wheat and clover, as well as on stands of poplar, sugar maple, red oak and white pine. Preliminary studies have found a reduction of 2-29 percent in the rate of growth of trees, and up to a 25 percent reduction in growth rate in one variety of wheat that appears to be particularly susceptible to ozone damage. The ozone appears to inhibit the process of photosynthesis in ways that are not yet fully understood.

World leaders are working together to preserve the upper atmospheric ozone layer that protects the entire planet from solar radiation. At the same time, leaders in affected local areas of individual nations must deal with the ozone resulting from human activities. This ozone, along with nitrogen oxides and sulphur dioxide, is part of the chemical soup of air pollutants that can cause harm to forests, croplands and people.

8. Toward a Strategy to Save Tropical Forests

Different articles in this book have cited many threats to the world's forest resources — acid rain, overcutting and overgrazing by the landless poor, deforestation due to timber harvesting and/or the conversion of forest land to human settlements, cattle ranches, roads, industrial developments, etc. Collectively the many benefits that people derive from forests have also been enumerated — help in maintaining the oxygen levels needed by all other living things for survival; timber for wood products; fuel; fruits, nuts and other foods; forage for wild and domesticated animals; places of recreation; habitats for other living things that provide food or enjoyment and fill important ecological niches. Gradually we are becoming more aware of the medical and genetic importance of the tree and other plant, animal and insect species that inhabit the world's forests, and of the roles that forested areas play in absorbing precipitation, reducing erosion and flooding, and maintaining ground and surface water levels during the drier seasons. And within recent years, a still further role — one of immense importance — is coming to be recognized: climate control, which helps to create and maintain rainfall patterns in downwind locations.

Nowhere are these tasks carried out in greater detail than in the world's tropical forests, which scientists estimate may shelter as many as half of the species of the world. Forty percent of the rain forests have already been destroyed or degraded. And if present trends are permitted to continue, the forests will all but disappear in two or three decades in many developing countries. That is why, in October 1985, a far-reaching five-year strategy to halt destruction of tropical forests (which lie in 56 Third World nations) was announced. Studies leading to the development of the strategy held the destruction of forests in Asia, Africa and Latin America responsible for much of the intensifying poverty throughout the Third World, noting that Ethiopia is enduring famine partly because land stripped of vegetation is now unfit for agricultural use.

The loss of tropical forests also affects world food production in another way. More than half of the 2 billion people worldwide who depend on firewood for cooking cut trees down faster than they can be replaced by natural processes. In areas where wood for fuel is scarce, an estimated 400 million tons of animal dung is burned each year, used as fuel by which to cook meals. The practice robs croplands of a major source of nutrients and is estimated to decrease world food harvests by 14 million tons per year. (To put the immensity of that loss in perspective, the United Nations estimated world grain shipments in 1985 at about 12 million tons.)

Recognizing the vulnerability to political instability engendered by increasing millions living in poverty, the land whose productivity is declining, and the accelerating disappearance of plant and animal species dependent upon the forested areas, Dr. T.H. Khoshoo, former secretary of the environment for India and a member of the task force that revealed the strategy, stated, "This is a classic example of a Third World problem the industrialized nations cannot afford to ignore."

The strategy, whose development was sponsored by the World Resources Institute of Washington, D.C., in collaboration with the United Nations Development Program and the World Bank, calls for an investment of $8 billion (double the present level) over the next five years in thousands of projects in conservation, training, research and tree planting. Andrew Maguire, vice president of the World Resources Institute said the plan "isn't just about saving trees, but entails an integrated approach to economic development, which preserves natural resources including forests and uses them wisely over time." Among the specific recommendations are:

• revision in each of the 56 countries in which tropical forests lie, of specified government policies (such as subsidies for large-scale cattle ranching or cheap timber contracts) that encourage exploitation, depletion or waste of forest resources;

• more intensive agricultural and rural development programs that could help the 250 million people already living within rain forests to establish sustainable farming systems so they will not encroach further on endangered forests;

• accelerated land-reform programs, and establishment of policies that encourage local participation in rural tree-planting programs and natural forest management;

• channeling future agricultural settlement into non-forest areas or suitable deforested areas;

• international support for development projects in transportation and irrigation expressly designed to avoid wasting or destroying forest resources.

The strategy is intended to supplement the ongoing efforts of the United Nation's Food and Agriculture Organization. It has also been presented to the United Nations World Commission on Environment and Development. The commission will consider whether to include the strategy in its report *Common Future*, due for presentation to the U.N. General Assembly in 1987.

9. Bringing National Policies in Line with Our Growing Understanding

Almost nowhere is the new puzzle of global interconnectedness and interdependence more challenging than in the realm of international relations. Gaining better understanding about the interrelationships among environmental and development-related problems is one thing. Translating that understanding into comprehensive proposals for action, like the one described for tropical rain forests, is the next step; but moving those recommendations into a political framework within which interactions with other nations are conducted is quite another thing. It means moving aside the criteria that have traditionally guided most international trade, aid and relations decisions — the desires to maintain or improve political relationships and/or protect or expand one 's own national economic interests — and looking at a much broader, more complex picture. It calls for a new political maturity that looks much more carefully at the interrelationships among individual societal problems and the implications of alternative solutions. It is a particularly difficult process in a nation like the United States, with its elected representatives who are often evaluated by how well they respond to noisy lobbists or protect the short-term interests of the particular mix of constituencies that comprises "back home."

Yet a quite exciting step toward this important political maturity is represented by the October 1985 report "A Congressional Agenda for Improved Resource and Environmental Management in the Third World: Helping Developing Countries Help Themselves." Prepared by a special task force of the Environmental and Energy Study Institute, an independent, non-partisan policy analysis organization that works closely with the Congressional Environmental and Energy Study Conference, the report lays out politically realistic and fiscally restrained recommendations for future congressional action. The task force, which had representatives from U.S. corporations, environmental and development organizations, and the Congress, concentrated on initiatives that would help developing countries help themselves.

The task force was initiated with the premise that "the long-term environmental, economic and political interests and national security of the United States depend far more than has been generally acknowledged on the success with which developing countries can wisely manage their resources for sustainable development. We believe that the only lasting way to achieve sound environmental and resource management in developing countries is to strengthen the capability of these countries to do so."

The interacting problems investigated during the development of the report included: management of dangerous chemicals; dependence upon expensive energy imports; pollution levels; the loss of forests and biological diversity; and land deterioration resulting from deforestation and desertification.

The report conclusions did not minimize the challenges that the world faces. It stated: "Much of the developing world is in the midst of an environmental and natural resources crisis which is undermining its drive for sustained economic growth . . . The problems of the poor countries generally make those of the United States and other rich countries pale by comparison. Moreover, because the struggling economies of the developing world generally are closely tied to the use of natural resources, the widespread deterioration of the renewable resource base has serious implications for the future, particularly when viewed together with rapidly expanding populations and the increasingly intricate economic and security interconnections among all nations."

Seeing U.S. involvement in finding solutions to these problems as critical, the report called upon the U.S. to provide global leadership, includingtaking action on 13 specific recommendations presented to Congress as "a package of interrelated and mutually supportive actions." The recommendations included the following:

• Provide long-term resource restoration assistance to Africa (not just emergency famine assistance).

• Establish a new Natural Resources Policy and Bureau within the U.S. Agency for International Development (to bring the United States' own bilateral development assistance into full alignment with the report's goals).

• Strengthen environmental authorities in developing countries (encouraging increases in their manpower, funding levels and clout).

• Build indigenous non-governmental organizations in developing countries (to help with broad public education and to keep officials' "feet to the fire").

• Improve environmental and related research and development and enhance related training (to increase the multiplier effects of progress in individual research or professional areas).

• Focus multilateral development bank attention on resource issues (assuring greater consistency among development and environmental protection efforts).

• Make optimal use of existing legislative mechanisms to achieve the above-stated goals (such as by expanding the existing Food for Peace program goals to include projects that conserve biological diversity).

Two additional recommendations that would seem to hold particular promise for the development of improved management of global resources are the recommendations to:

• Incorporate natural resource assessments into project cost/ benefit analysis, and

• Review options for using foreign debt to encourage sustainable development.

It is perhaps through thinking like this last idea that one of the most important "missing pieces" of our new global puzzle may be found — the lack of the current global economic "accounting system" to reflect the value of the planetary life-support processes that we have traditionally counted as "free services" — the services that include protection of Earth from the sun's radiation; regulation of climate and weather; recycling of nutrients; disposal of many waste products; replenishment of soils; pollenation; natural control of many pest and disease vectors; and maintenance of the genetic bank, the source of all our present and future crops and food.

The foreign debt recommendation of the EESI task forces proposes possible cancellation or rescheduling of foreign debt in exchange for developing country efforts to undertake environmental and natural resource projects which will improve the long-term prospects for sustained growth and debt repayment. This concept opens the way, as well, for even broader, international agreements covering nations whose resources (like U.S. and Canadian prime agricultural lands and the developing nations' tropical forests) play key roles in supporting human life on Earth. Such agreements could establish some system to encourage and reward these nations for sound stewardship of these globally important resources which are the responsibility of the individual nations through historic accident alone.

As Gus Speth, president of the World Resources Institute, and chairman of the EESI task force, stated: "These newer environmental concerns transcend borders, national laws and local customs. As a result, the politics needed to meet present and future challenges require a new vision and a new diplomacy, new leaders and new coalitions."

10. Retaining Our Willingness to Risk

This final example is different from any of the others that precede it. It does deal with a very contemporary global environmental issue — the problem of how to provide people and communities with electric energy while minimizing environmental damage and holding economic costs to a reasonable level. The difference lies in the fact that this shared experience took place in the late 1800s. It comes from a guest editorial written for The Environmentalist *by Thomas A. Brosnan, who for many, many years was a major figure within the electrical energy generation field in the eastern United States. His story tells us of the immense challenges that were involved in the effort to make electricity widely available in this section of the country. It is also a reminder that when determined and creative people take risks, agree to use the very best expertise available to avoid problems, and commit to cooperate across national boundaries, long-standing constraints on human living conditions can be overcome. The time period within which this particular issue was brought from idea to reality is also worth noting. The situation offers a marked contrast to what is possible today. (Published in 1982 in Volume 2, Issue 1 of* The Environmentalist *by Thomas A. Brosnan.)*

Every project through which energy is converted from its basic forms into electric energy involves an interesting relationship with environment and economics, but I place at the top of the list the interplay of these considerations at Niagara Falls, which is a portion of the boundary between New York State and Canada's Province of Ontario.

Practically the entire water outflow from four of the North American Great Lakes (Superior, Huron, Michigan and Erie) passes through the Niagara River into Lake Ontario and the St. Lawrence River on its way to the Atlantic Ocean. The plunge of the river from the elevation of Lake Erie to that of Lake Ontario has created the environmental beauty of Niagara Falls along with its spectacular rapids and whirlpools. The environmental importance of the area goes beyond its beauty. The lower gorge has been formed through the eons as the falling waters have cut a slot through the rock of the escarpment. That gorge reveals not only the layered rock formations but shows how those formations have permitted the falls to continue as a sheer drop rather than changing to sloping rapids. It also reveals how the source of the river was changed by the deposits of the ice age.

From the time of the early settlers there was not only appreciation of the beauty of Niagara Falls, but also recognition of the energy potential in the falling waters. In the middle of the 19th century, various projects were initiated to divert water from the upper river through ditches and pipes and into small waterwheels in order to provide mechanical energy to operate equipment in mills and factories. The most ambitious project was the construction, in the 1860s, of a canal from which water was offered for sale, as a power source, to those who might be interested in constructing mills along the canal.

The limited availability of dependable, small-scale hydraulic equipment and the financial risks related to such installations were major deterrents to the expansion of this arrangement for energy distribution and conversion. It was realized also that it made very limited use of the potential energy in the waters above the falls and had some serious environmental drawbacks. This led a group, under the leadership of Edward Dean Adams, to explore the merits of a central station for the conversion of the basic energy into a form more readily adaptable for distribution to and use in manufacturing and commercial establishments. What challenges this presented!

It was recognized early in the planning that the economies-of-scale dictated a large-size installation and it was decided to design a plant with ten 5,000 HU units. This would require the installation of much larger hydraulic facilities than had previously been designed.

Into what form of energy should the mechanical energy in the hydraulic turbines be converted? Reports indicate that the early thinking favored conversion to compressed air for distribution and utilization. There were some who urged consideration of direct-current electricity. The use of this form of energy distribution had been initiated only a few years earlier with the construction of Thomas Edison's Pearl Street generating station in New York City. Both the compressed air proposal and the direct-current electrical system would impose serious technical and economic limitations on the distance the energy could be delivered from the central stations to prospective utilization locations.

A small group suggested the introduction of alternative-current electricity. The equipment and system concepts for this approach were barely out of the research status at that time. However, the pioneering design and development work had been done on transformers, induction motors, rotary converters and other equipment. This meant that this system, which held promise for major expansion of the distances for economical energy transmission, was at the threshold of commercial implementation, if someone was prepared to assume the risks.

In an effort to get the best guidance for these monumental decisions, the opinions of the best-known engineers and scientists in the United States were solicited. In addition, discussions were held with knowledgeable people in other countries. This led to setting up the International Niagara Commission in June 1890, under the chairmanship of Sir William Thomson of Great Britain. Of the other distinguished personnel on the commission, one each was from Switzerland, France, United States and England.

By May 1893 the opinions and recommendations of

the experts were available, and the directors of the construction company made the decision that the energy conversion would be to alternative-current electricity. By August 1895 the first generating units were in operation. In November 1896 the transmission of electric energy was initiated along the Niagara frontier from Niagara to Buffalo. With that event there came into existence the alternative-current electrical system of generation, transmission and distribution that has become a keystone of our way of life throughout the world today.

Throughout this development and during the decades that followed, the environmental attractiveness of the Niagara area has been carefully safeguarded. Regulation has played an important part in this achievement. Since the geographical boundary between the United States and Canada is located in the Niagara River, regulatory responsibility resides in the coordinated action of two nations. What has been accomplished is a tribute to the exercise of reason and cooperation. During the earlier years the permits for water diversion from the upper river were based on fixed amounts for specified purposes. More recently this regulatory arrangement has been turned around to specify what water flow must be maintained over the falls at various times. One amount is stipulated during the daylight hours and a smaller amount at night. The mandated flow is also higher in summer than during the winter months.

This method of regulation clearly established the priority of the enjoyment of the beauty of the falls. It also recognized the importance of preservation of that beauty for future generations by minimizing rock erosion at the falls. This has been further implemented by the installation of remedial works to improve the flow pattern of the river above the falls. The regulatory arrangement has tremendous economic merit in that it permits maximum conversion of the available potential energy under varying conditions of river flow and with adequate recognition of environmental considerations.

In summary, we observe at Niagara Falls an interesting illustration of the controlled interrelationship of environment, energy conversion and economics. The technical and economic risks involved were monumental. The procedures that were followed made use of the techniques and projects that had just become available from worldwide research and design. There were major differences of opinion among renowned experts so that careful reasoning was needed in making decisions. The international regulations that were propounded improved and helped the preservation of the environmental attractiveness of the area and still gave suitable recognition to the economic importance of the potential energy in the waters above the falls.

PART FOUR

Conclusion

We can see from the examples just reviewed that individual elements of the needed overall effort addressing global environmental problems are coming together:

• The data needed for improved decisions regarding sustainable economic development and the management of environmental systems is expanding.

• Beginnings toward improved management of toxic substances are being made through pilot projects and new cooperation among industry, researchers and governmental institutions.

• Efforts such as UNEP's Regional Seas Program and the CITES program are giving us practical experience with multinational cooperative management of shared natural resources.

• Emerging techniques like conflict resolution and consensus building hold promise for finding ways other than destructive riots, strikes and military aggression for dealing with differences of opinion.

• New ideas on how to protect global natural resources — the foundation of the world economy — and at the same time stimulate healthy economic growth in the Third World are coming forth. Norman Myers, general editor of *GAIA: An Atlas for Planetary Management*, has suggested compensatory payments to promote world conservation, including surcharges on purchases of materials derived from threatened areas, with revenues to go to a conservation superfund, and relaxation of trade quotas to encourage Third World economic growth. The recently released *"Congressional Agenda for Improved Resource and Environmental Management in the Third*

World: Helping Developing Countries Help Themselves," likewise recommends consideration of canceling or delaying international debts in exchange for development of improved environmental management within Third World nations.

• The World Conservation Strategy provides an overall, if preliminary, blueprint for balancing economic growth and the needs of environmental systems. Its goal of sustainable development is being brought closer to a working reality through the preparation of strategies for a number of individual nations.

• And the World Commission on Environment and Development is an existing channel into which all the improving data, clarification of needs and problems, and ideas for replicable solutions can be fed as the world seeks better ways to address these interrelated environmental problems and their economic, social and political implications.

Citizens of the United States can take justifiable pride in the role their country has played thus far in adjusting to this puzzling new world of interdependence. Visionary American conservationists, like Hal Coolidge, helped set up the International Union for the Conservation of Nature and Natural Resources and the World Wildlife Fund, and pressed to hold the Stockholm Conference. The United States led the way toward heading off resource conflicts in the Antarctic by urging the ratification of the Antarctic Treaty of 1959. Leaders from this country encouraged the establishment of the World Heritage Convention, the Convention on International Trade in Endangered Species of Fauna and Flora, and many of the other landmark international agreements. This nation's research and policy centers have helped spell out the dimensions of resource and environmental systems problems, and identify potential solutions. Our extensive environmental legislation has often provided a guide for comparable efforts in other parts of the world. Study of the resources and widely varied climate and vegetative zones of this vast nation has given insights into the ways that a number of key global natural systems function. And our immense good fortune at being relatively free from internal stress or external conflict has permitted funds, manpower, time and public attention to be devoted to the examination of these issues.

These individual steps are producing results in scattered places and among individual elements of the total problem; and yet neither in the United States nor the world at large is the progress nearly fast or comprehensive enough. Steeped in the perception of individual nations as independent, politically defined entities, it is hard to accept the growing evidence that each nation, in addition to practicing sound environmental management itself, must rely on other nations, international bodies and multinational corporations to protect the natural systems, cycles and processes upon which life in each nation depends. Equally hard to accept is the rising evidence that nations, whose interactions have often been based on competition and conflict, must cooperate if we are to wisely manage the resources upon which the economies of every nation have come to ride.

Flat to round; independent to interdependent

Hundreds of years ago, when the realization that the world was round, not flat, began to take hold in human consciousness, there must have been great fear and trepidation. Those in positions of influence and power based on a "flat" world must have felt terribly threatened; while those afraid of change in any form must have been devastated by the new concept. Yet gradually, as the evidence of this newly recognized eternal truth became impossible to deny, the painstaking tasks of readjusting thinking, governments and societies took place — and the immense benefits of that formerly unrecognized reality were finally able to flow into human lives. For it was the well-being of people that was improved by this shift in human perceptions and behaviors; no one had ever had to tell the oceans of the world, the winds, the whales or all the migratory species how the planet was shaped.

Increasing interdependence among the nations of the world has brought us to a similar major choice point in human consciousness — and one that may require just as great changes in behavior as when people finally accepted that the world was round. We cannot let fear of these changes slow the behavioral shifts that this new reality calls for, for the need to accept and adjust to this new perception is much more pressing than when that earlier historic shift in human understanding took place.

What was lost then by reluctance to accept the reality that the Earth is round were chances to widen and extend human options — but those opportunities remained until the human family on Earth was ready to take advantage of them. In contrast, what is being lost now, through delay in responding to interdependence, is options — leaving us with less time to heal weakened natural support systems; less time to extend the lifetime of limited resources; less time to adjust our economic, investment and aid programs; and less time to find sustainable ways to meet the needs of a growing global, human population.

For, once again, it is principally human beings who will benefit from this second major revision in our perception of the planet. Other forms of life seem able to adjust to almost any conditions — as indicated by the plant growth, insects and small animals returning to the area devastated by the volcanic eruption of Mount St. Helens. It is people who are most at risk from continued deterioration of global environment. With our constant need for clean air and water, uncontaminated food, shelter, medical care and other resources, we will suffer most if planetary life support systems are permitted to fail, and it is we who will benefit if we let the new reality of interdependence among nations and collective dependence of all nations upon natural systems guide us toward, as Bucky Fuller would say, "perceptions and behaviors that do work" within this puzzling new world.

A key role for Americans

For Americans, who are fortunate enough to enjoy the greatest freedom and one of the highest living standards in the world, what is at stake in terms of potential loss in life quality if the adjustment is not made is immense. It is very much in our own best interests to work toward the resolution of environmental problems — whether local, in which fellow worker safety or public health is at stake, or global, in which acid rain, toxic pollution, deforestation, desertification or soil erosion are threatening the natural systems which support life in this nation and everywhere else on the planet.

If Americans accept a leadership role in this transition it will not be the first time they have led the world into a new era. The people who have settled in America have helped the world expand and redefine what "freedom" and "individual opportunity" can mean. Now, the opportunity exists to help the world redefine this newest global reality — "interdependence" — seeking ways not to weaken sovereign rights but to affirm sovereign responsibilities and create synergistic relationships among nations that will lead to new partnerships among themselves and with nature.

One of Americans' biggest assets in leading the way toward this vitally important transition is that we are a society open to self-examination and change. Being free of the internal suffering and turmoil that wracks much of the Third World, and unhampered by too strong ties to the past, we are in a position to help identify which cultural, political, economic and environmental perceptions and behaviors are and which are not any longer appropriate in this world of increasing global interdependence. As a compassionate and caring people, we can help reveal this new global puzzle of interdependence for what it is — a major choice point for humanity. And we can inspire and encourage the development of appropriate new behaviors so that, once again, the human population on the planet can benefit from this latest evolutionary leap in human understanding. We can put the pieces of this new global puzzle together.

"If the world is going to be changed it will be because a new image appears that people will hold to and in which they will place their beliefs and their hopes."

Sacred Influences
J.G. Bennett

Appendix 1:
Abbreviations Frequently Encountered in the International Environmental Arena.

ACIC	- American Committee for International Conservation
AID	- U.S. Agency for International Development
CITES	- Convention on the International Trade in Endangered Species of Flora and Fauna
EPA	- U.S. Environmental Protection Agency
FAO	- Food and Agricultural Organization of the United Nations
GEMS	- Global Environmental Monitoring System/UNEP Earthwatch Program
GRID	- Global Resources Information Database/UNEP GEMS Program
IIED	- International Institute for Environment and Development
IRPTC	- International Registry of Potentially Toxic Chemicals/UNEP
IUCN	- International Union for the Conservation of Nature and Natural Resources
IEEP	- UNESCO/UNEP International Environmental Education Program
MAB	- Man And the Biosphere Program/UNESCO
NASA	- U.S. National Aeronautic and Space Administration
OECD	- Organization for Economic Co-operation and Development
OEOA	- United Nations Office of Emergency Operations in Africa
UNDP	- United Nations Development Program
UNEP	- United Nations Environment Program
UNESCO	- United Nations Educational, Scientific and Cultural Organization
USDA	- U.S. Department of Agriculture
USGS	- U.S. Geological Survey/Department of the Interior
WCED	- World Commission on Environment and Development
WFF	- World Wildlife Fund/International
WHO	- World Health Organization
WMO	- World Meteorological Organization
WICEM	- World Industry Conference on Environmental Management

Appendix 2: Publications Referenced on Global Environmental, Economic, Social and Political Issues

Brown, Lester. *State of the World 1984, State of the World 1985*. Washington, D.C.: Worldwatch Institute, 1984, 1985.

Conservation Foundation. *State of the Environment 1984*. Washington, D.C., 1984.

Ehrlich, Paul, Carl Sagan, Donald Kennedy and Walter Orr Roberts. *The Cold and the Dark: The World After Nuclear War*. New York: Norton, 1984.

Environmental and Energy Study Institute. *A Congressional Agenda for Improved Resource and Environmental Management in the Third World: Helping Developing Nations Help Themselves*. Washington, D.C., 1985

Global Future: Time to Act. Report to the President on Global Resources, Environment, and Population. Council on Environmental Quality and U.S. Department of State. Washington, D.C.: Government Printing Office, January 1981.

Global 2000 Report to the President. Gerald O. Barney, Study Director. Council on Environmental Quality and Department of State. Three volumes. Volume I: *Entering the Twenty-first Century* is a summary report. Washington, D.C.: Government Printing Office, 1980.

Institute of Ecotechnics. *The Biosphere Catalogue*. London: Synergetic Press. 1985.

Meadows, D. H., D. L. Meadows, J. Randers and W. Behrens. *The Limits to Growth: A Report for the Club of Rome's Project on the Predicament of Mankind*. Signet, New American Library, 1972.

Meadows, Dennis L. *Alternatives to Growth: A Search for Sustainable Futures*. Cambridge, MA: Ballinger, 1977.

Myers, Norman, Gen. Ed. *GAIA: An Atlas of Planetary Management*. Garden City, NY: Anchor Books, 1984.

North-South: A Program for Survival. Independent Commission on International Development Issues (The Brandt Commission). Cambridge, MA: MIT Press, 1980.

Simon, Julian, Hermann Kahn. *The Resourceful Earth: A Response to Global 2000*. New York: Bazil Blackwell, 1984.

United Nations Environment Program *The State of the Environment 1972-1982*. Nairobi, Kenya, 1982.

World Resources Institute. *The Global Possible: Resources, Development and the New Century*. Washington, D.C., 1984.

World Conservation Strategy: Living Resource Conservation for Sustainable Development. Gland, Switzerland: International Union for the Conservation of Nature and Natural Resources, 1980. Available through Unipub, P.O. Box 433, Murray Hill Station, New York, NY 10016.

World Industry Conference on Environmental Management (WICEM): Outcome and Reaction. Paris: United Nations Environment Program, 1984. Available through the Industry and Environment Office, UNEP, 17 rue Margueritte, 75017, Paris, France.

Appendix 3: Institutions and Projects Referenced in the Text

Agency for International Development
Bureau for External Affairs
Washington, D.C. 20523 USA

American Forestry Association
1319 18th Street, N.W.
Washington, D.C. 20036 USA

Center for Environmental Management
Dr. Anthony Cortese, Director
18 Latin Way
Tufts University
Medford, MA 02154 USA

Earthscan Bulletin
3 Endsleigh Street
London WC1H ODD United Kingdom
(U.S. Office:)
c/o IIED
1717 Massachusetts Avenue
Washington, D.C. 20036 USA

Earthwatch Global Monitoring Program - see UNEP

Environmental Protection Agency,
U.S. Office of Public Affairs (A-107)
Washington, D.C. 20460 USA

Environmental and Energy Study Institute
410 First Street, S.E.
Washington, D.C. 20003 USA

Food and Agricultural Organization of the U.N.
1776 F Street, N.W.
Washington, D.C. 20437 USA

Global Tomorrow Coalition
1325 G Street, N.W.
Suite 1003
Washington, D.C. 20005 USA

International Institute for Environment & Development
1717 Massachusetts Ave.
Washington, D.C. 20036 USA

International Union for the Conservation
of Nature and Natural Resources
1196 Gland
Switzerland

Institute of Ecotechnics
24 Old Gloucester Street
London, WC1 3AL
United Kingdom

Man and the Biosphere Program
UNESCO
7 Place de Fontenoy
Paris France

Massachusetts Audubon Society
Lincoln, Massachusetts 01703 USA

Myers, Dr. Norman
Upper Meadow
Old Road, Headington
Oxford, OX3 852
United Kingdom

National Aeronautic and Space Administration, U.S.
Public Affairs Office
Washington, D.C. 20546 USA

New England Governors' Conference, Inc.
Charles Fausold, Director of Environmental Programs
76 Summer Street
Boston, MA 02110 USA

Office of Emergency Operations in Africa
United Nations
New York, NY 10017 USA

The Conservation Foundation
1255 23rd Street, N.W.
Washington, D.C. 20037 USA

The Environmentalist
Science and Technology Letters Limited
12 Clarence Road
Kew, Surrey TW9 3NK
United Kingdom

The Pennsylvania State University
New Kensington Campus
3550 Seventy Street Road
New Kensington, PA 15068 USA

United Nations Development Program (UNDP)
United Nations
New York, NY 10017 USA

United Nations Educational, Scientific and Cultural Organization (UNESCO) UNESCO/UNEP International Environmental Education Program
7 Place de Fontenoy
Paris, France

United Nations Environment Program (UNEP)
P.O. Box 30552
Nairobi, Kenya
(addresses for information in the U.S.:)

UNEP
1889 F Street,
N.W. Washington, D.C. 20006 USA
Joan Martin Brown, Washington Senior Liaison
UNEP United Nations,
NY, NY 10017 USA
Dr. Noel J. Brown North American Representative

University-Industry Cooperative Center for Research in Toxics and Hazardous Materials (State of New Jersey)
Public Policy & Education Division,
Dr. Michael Greenberg, Director
Lucy Stone Hall, Kilmer Campus
Rutgers University
New Brunswick, NJ 08903 USA

World Bank
1818 H Street, N.W.
Washington, D.C. 20433 USA

World Commission on Environment and Development
William Ruckelshaus, Member
c/o The Conservation Foundation
1255 23rd Street
Washington, D.C. 20037 USA
Maurice F. Strong, Member
355 Burrard Street
Vancouver, B.C. V6C 2G8 Canada

World Heritage Convention
UNESCO
7 Place de Fontenoy
Paris, France

World Resources Institute
1735 New York Avenue, N.W.
Washington, D.C. 20006 USA

Worldwatch Institute 1776 Massachusetts Ave., N.W.
Washington, D.C. 20036 USA

World Wildlife Fund International
1196 Gland
Switzerland

World Wildlife Fund/U.S.
1255 23rd Street, N.W.
Washington, D.C. 20037 USA

INDEX